Table of Contents

3 ROBOTIC SENSING

4 DEGREES OF FREEDOM MECHANICS

5 THREE LAWS OF ROBOTICS

6 MOBILE ROBOTS

9 CLOUD ROBOTICS

List of Abbreviations

AMF	Additive Manufacturing File
AGVs	Automated Guided Vehicles
FPS	First-Person Shooter
FDM	Fused Deposition Modeling
GMO	Genetically Modified Organism
CCD	Charge-Coupled Device
HRI/OS	Human-Robot Interaction Operating System
CSCC	Complex Spectrum Circle Centroid
CAD	Computer Aided Design
DMLS	Direct Metal Laser Sintering
DOF	Degree of Freedom
LOM	Laminated Object Manufacturing
SNPs	Single Nucleotide Polymorphisms
6DoF	Six Degrees of Freedom
PC	Personal Computer
PCR	Polymerase Chain Reaction
RFID	Radio Frequency Identification
RAT	Robot Assisted Therapy
UAVs	Unmanned Aerial Vehicles
SAR	Search and Rescue
SLM	Selective Laser Melting
SLS	Selective Laser Sintering
WHB	Wolfram Bergbau und Hütten AG

Preface

Robotics can be described as the current pinnacle of technical development. Robotics is a confluence science using the continuing advancements of mechanical engineering, material science, sensor fabrication, manufacturing techniques, and advanced algorithms. The study and practice of robotics will expose a dabbler or professional to hundreds of different avenues of study. For some, the romanticism of robotics brings forth an almost magical curiosity of the world leading to creation of amazing machines. A journey of a lifetime awaits in robotics. Robots in antiquity and through the Middle Ages were used primarily for entertainment. However, the 20th century featured a boom in the development of industrial robots. Through the rest of the century, robots changed the structure of society and allowed for safer conditions for labor. Robots have also been influential in the media and profitable for toy manufacturers. Industry has benefited drastically from the expanse of a robotic work force. Automated machines have taken over the duties of dangerous and mundane jobs from humans, allowing greater productivity. Because robots never tire, extra shifts have been added to factories. Just as industrial robots are remaking the manufacturing industry by creating higher production rates and improved quality, and revolutionizing the way we think about and administer business processes, IT support processes, workflow processes, remote infrastructure and back-office work. Robots' main position in society are in a capacity to assist humans by taking on the jobs that are dirty, dull or dangerous.

This Text provides accessible introduction to robotics that spans the most basic concepts and the most novel applications; for students, teachers and professionals.

Content Coverage

Chapters One and *Two* present the history of robots and robot kinematics, correspondingly. In the early 20th century, the notion of a humanoid machine was developed. Today, it is now possible to envisage human sized robots with the capacity for near human thoughts and movement.

Chapters Three and *Four* present a coverage of robotic sensing and degree of freedom, consistently. Robotic sensing mainly gives robots the ability to see, touch, hear and move and uses algorithms that require environmental feedback, however the degree of freedom (DOF) of a mechanical system is the number of independent parameters that define its configuration.

Chapters Five and *Six* are intended to focus on three laws of robotics and mobile robots, separately. Mobile robots have the capability to move around in their environment and are not fixed to one physical location.

Chapters Seven and *Eight* clarify the fundamentals of BEAM robotics and Nano robotics techniques, separately. BEAM robotics is a style of robotics that primarily uses simple analogue circuits, such as comparators, instead of a microprocessor in order to produce an unusually simple design.

Chapter Nine is cloud robotics that attempts to invoke cloud technologies such as cloud computing, cloud storage, and other Internet technologies centered on the benefits of converged infrastructure and shared services for robotics.

CHAPTER 1

History of Robots

Chapter Contents

- **Introduction**
- **Early Legends**
- **Early Beginnings**
- **Remote-Controlled Systems**
- **Humanoid Robots**
- **Modern Autonomous Robots**

The word robot is usually used to refer to a mechanical agent that performs one or more tasks in which it mimics a human or animal agent either through programming or commands. Another word used synonymously is automaton. Virtual robots exist, but are most often called by the abbreviated name bots.

INTRODUCTION

Thc history of robots has its origins on the ancient world. The modern concept began to be developed with the onset of the Industrial Revolution which allowed for the use of complex mechanics and the subsequent introduction of electricity. This made it possible to power machines with small compact motors. In the early 20th century, the notion of a humanoid machine was developed. Today, it is now possible to envisage human sized robots with the capacity for near human thoughts and movement.

The first uses of modern robots were in factories as industrial robots – simple fixed machines capable of manufacturing tasks which allowed production without the need for human assistance. Digitally controlled industrial robots and robots making use of artificial intelligence have been built since the 1960s.

EARLY LEGENDS

Hephaestus, Greek god of craftsmen.

Concept of artificial servants and companions date at least as far back as the ancient legends of Cadmus, who sowed dragon teeth that turned into soldiers, and the myth of Pygmalion whose statue of Galatea came to life. Many ancient mythologies included artificial people, such as the talking mechanical handmaidens built by the Greek god Hephaestus (Vulcan to the Romans) out of gold, the clay golems of Jewish legend and clay giants of Norse legend. Chinese legend relates that in the 10th century BC, Yan Shi made an automaton resembling a human in an account from the *Lie Zi* text.

In Greek mythology, Hephaestus created utilitarian three-legged tables that could move about under their own power and a bronze man, Talos, that defended Crete. Talos was eventually destroyed by Media who cast a lightning bolt at his single vein of lead. To take the golden fleece Jason was also required to tame two fire breathing bulls with bronze hooves; and like Cadmus he sowed the teeth of a dragon into soldiers.

The Indian *Lokapannatti* (11th/12th century) tells the story of King Ajatashatru of Magadha who gathered the Buddhas relics and hid them in an underground stupa. The Buddhas relics were protected by mechanical robots (bhuta vahana yanta), from the kingdom of Roma visaya; until they were disarmed by King Ashoka. In the Egyptian legend of Rocail, the younger brother of Seth created a palace and a sepulcher containing autonomous statues that lived out the lives of men so realistically they were mistaken for having souls. ... In Christian legend, several of the men associated with the introduction of Arabic learning (and, through it, the reintroduction of Aristotle and Hero's works) to medieval Europe devised brazen heads that could answer questions posed to them. Albertus Magnus was supposed to have constructed an entire android who could perform some domestic tasks but was destroyed by Albert's student Thomas Aquinas for disturbing his thought. The most famous legend concerned a bronze head devised by Roger Bacon which was destroyed or scrapped after he missed its moment of operation.

Automata were popular in the imaginary worlds of medieval literature. For instance, the Middle Dutch tale *Roman van Walewein* ("The Romance of Walewein, early 13th century) described mechanical birds and angels producing sound by means of systems of pipes.

EARLY BEGINNINGS

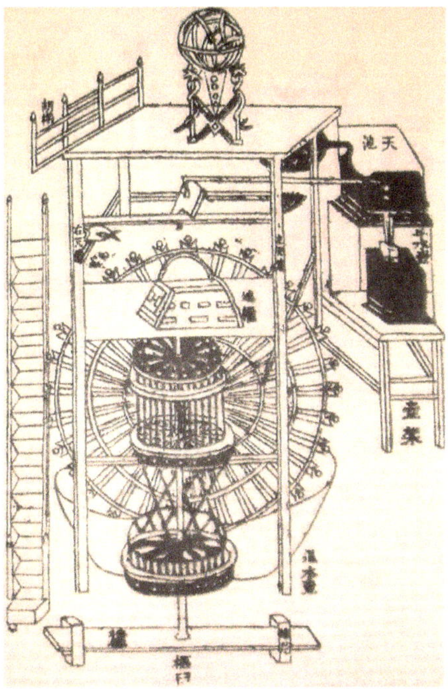

The water-powered mechanism of Su Song's astronomical clock tower, featuring a clepsydra tank, waterwheel, escapement mechanism, and chain drive to power an armillary sphere and 113 striking clock jacks to sound the hours and to display informative plaques.

Concepts akin to a robot can be found as long ago as the 4th century BC, when the Greek mathematician Archytas of Tarentum postulated a mechanical bird he called "The Pigeon" which was propelled by steam. Yet another early automaton was the clepsydra, made in 250 BC by Ctesibius of Alexandria, a physicist and inventor from Ptolemaic Egypt. Hero of Alexandria (10–70 AD) made numerous innovations in the field of automata, including one that allegedly could speak.

Taking up the earlier reference in Homer's Iliad, Aristotle speculated in his *Politics* (ca. 322 BC, book 1, part 4) that automatons could someday bring about human equality by making possible the abolition of slavery: – There is only one condition in which we can imagine managers not needing subordinates, and masters not needing slaves. This condition would be that each instrument could do its own work, at the word of command or by intelligent anticipation, like the statues of Daedalus or the tripods made by Hephaestus, of which Homer relates that Of their own motion they entered the conclave of Gods on Olympus, as if a shuttle should weave of itself, and a plectrum should do its own harp playing.

In ancient China, an account on automata is found in the *Lie Zi* text, written in the 3rd century BC, in which King Mu of Zhou (1023–957 BC) is presented with a life-size, human-shaped mechanical figure by Yan Shi, an "artificer".

The Cosmic Engine, a 10-metre (33 ft) clock tower built by Su Song in Kaifeng, China, in 1088, featured mechanical mannequins that chimed the hours, ringing gongs or bells among other devices.

Al-Jazari's programmable humanoid robots.

Al-Jazari (1136–1206), a Muslim inventor during the Artuqid dynasty, designed and constructed a number of automatic machines, including kitchen appliances and musical automata powered by water. One particularly complex automaton included four automatic musicians that floated on a lake.

Hero's works on automata were translated into Latin amid the 12th century Renaissance. The early 13th-century artist-engineer Villard de Honnecourt sketched plans for several automata. At the end of the thirteenth century, Robert II, Count of Artois, built a pleasure garden at his castle at Hesdin that incorporated a number of robots, humanoid and animal.

One of the first recorded designs of a humanoid robot was made by Leonardo da Vinci (1452–1519) in around 1495. Da Vinci's notebooks, rediscovered in the 1950s, contain detailed drawings of a mechanical knight in armour which was able to sit up, wave its arms and move its head and jaw. The design is likely to be based on his anatomical research recorded in the *Vitruvian Man* but it is not known whether he attempted to build the robot (see: Leonardo's robot). In 1533, Johannes Müller von Königsberg created an automaton eagle and fly made of iron; both could fly. John Dee is also known for creating a wooden beetle, capable of flying.

Tea-serving karakuri, with mechanism, 19th century. Tokyo National Science Museum.

Around 1700, many automatons were built including ones capable of acting, drawing, flying, and playing music; some of the most famous works of the period were created by Jacques de Vaucanson in 1737, including an automaton flute player, tambourine player, and his most famous work, "The Digesting Duck". Vaucanson's duck was powered by weights and was capable of imitating a real duck by flapping its wings (over 400 parts were in each of the wings alone), eat grain, digest it, and defecate by excreting matter stored in a hidden compartment.

The Japanese craftsman Hisashige Tanaka, known as Japan's Edison, created an array of extremely complex mechanical toys, some of which were capable of serving tea, firing arrows drawn from a quiver, or even painting a Japanese *kanji* character. The landmark text *Karakuri Zui* (*Illustrated Machinery*) was published in 1796.

REMOTE-CONTROLLED SYSTEMS

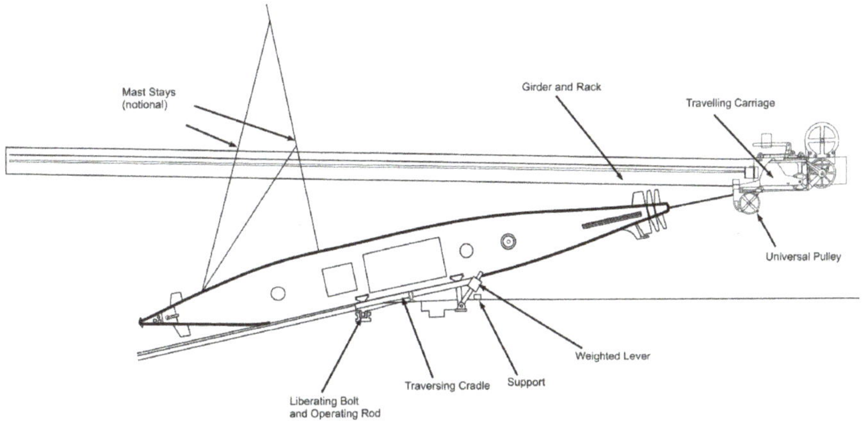

The Brennan torpedo, one of the earliest "guided missiles".

Remotely operated vehicles were demonstrated in the late 19th century in the form of several types of remotely controlled torpedoes. The early 1870s saw remotely controlled torpedoes by John Ericsson (pneumatic), John Louis Lay (electric wire guided), and Victor von Scheliha (electric wire guided).

The Brennan torpedo, invented by Louis Brennan in 1877 was powered by two contra-rotating propellers that were spun by rapidly pulling out wires from drums wound inside the torpedo. Differential speed on the wires connected to the shore station allowed the torpedo to be guided to its target, making it "the world's first *practical* guided missile". In 1898 Nikola Tesla publicly demonstrated a "wireless" radio-controlled torpedo that he hoped to sell to the U.S. Navy.

Archibald Low was known as the father of radio guidance systems for his pioneering work on guided rockets and planes during the First World War. In 1917, he demonstrated a remote controlled aircraft to the Royal Flying Corps and in the same year built the first wire-guided rocket. In the winter of 1970, the Soviet Union explored the surface of the moon with the lunar vehicle Lunokhod 1, the first roving remote-controlled robot to land on another celestial body.

HUMANOID ROBOTS

The term "robot" was first used to denote fictional automata in the 1921 play *R.U.R.* (Rossum's Universal Robots) by the Czech writer, Karel Čapek. According to Čapek, the word was created by his brother Josef from the Czech robota, meaning servitude. The play, *R.U.R,* replaced the popular use of the word "automaton" with the word "robot." In 1927, Fritz Lang's Metropolis was released; the Maschinenmensch ("machine-human"), a gynoid humanoid robot, also called "Parody", "Futura, Robotrix, or the Maria impersonator (played by German actress Brigitte Helm), was the first robot ever to be depicted on film. In many films, radio and television programs of the 1950s and before, the word "robot" was usually pronounced "robit," even though it was spelled "bot" and not "bit." Examples include "The Lonely" episode of the TV series "The Twilight Zone," first aired on November 15, 1959, and all episodes of the sci-fi radio program "X Minus One."

Many robots were constructed before the dawn of computer-controlled servomechanisms, for the public relations purposes of major firms. These were essentially machines that could perform a few stunts, like the automatons of the 18th century. In 1928, one of the first humanoid robots was exhibited at the annual exhibition of the Model Engineers Society in London. Invented by W. H. Richards, the robot Eric's frame consisted of an aluminium body of armour with eleven electromagnets and one motor powered by a twelve-volt power source. The robot could move its hands and head and could be controlled through remote control or voice control.

The first humanoid robot was a soldier with a trumpet, made in 1810 by Friedrich Kaufmann in Dresden, Germany. The robot was on display until at least April 30, 1950.

Westinghouse Electric Corporation built Televox in 1926 – it was a cardboard cutout connected to various devices which users could turn on and off. In 1939, the humanoid robot known as Elektro was debuted at the World's Fair. Seven feet tall (2.1 m) and weighing 265 pounds (120.2 kg), it could walk by voice command, speak about 700 words (using a 78-rpm record player), smoke cigarettes, blow up balloons, and move its head and arms. The body consisted of a steel gear cam and motor skeleton covered by an aluminum skin. In 1928, Japan's first robot, Gakutensoku, was designed and constructed by biologist Makoto Nishimura.

MODERN AUTONOMOUS ROBOTS

In 1941 and 1942, Isaac Asimov formulated the Three Laws of Robotics, and in the process of doing so, coined the word "robotics". In 1948, Norbert Wiener formulated the principles of cybernetics, the basis of practical robotics.

The first electronic autonomous robots with complex behaviour were created by William Grey Walter of the Burden Neurological Institute at Bristol, England in 1948 and 1949. He wanted to prove that rich connections between a small number of brain cells could give rise to very complex behaviors - essentially that the secret of how the brain worked lay in how it was wired up. His first robots, named *Elmer* and *Elsie*, were constructed between 1948 and 1949 and were often described as *tortoises* due to their shape and slow rate of movement. The three-wheeled tortoise robots were capable of phototaxis, by which they could find their way to a recharging station when they ran low on battery power. Walter stressed the importance of using purely analogue electronics to simulate brain processes at a time when his contemporaries such as Alan Turing and John von Neumann were all turning towards a view of mental processes in terms of digital computation. His work inspired subsequent generations of robotics researchers such as Rodney Brooks, Hans Moravec and Mark Tilden. Modern incarnations of Walter's *turtles* may be found in the form of BEAM robotics.

History of robots is linked to the development of artificial intelligence.

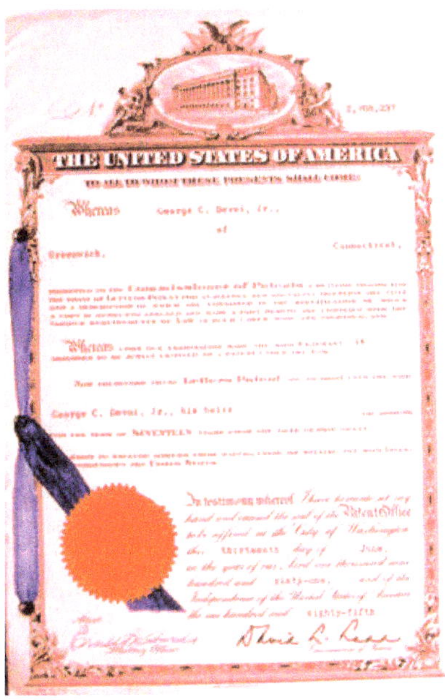
THE UNITED STATES OF AMERICA

Whereas George C. Devol, Jr., of Greenwich, Connecticut,

George C. Devol, Jr., his heirs

U.S. Patent 2,988,237, issued in 1961 to Devol.

The first digitally operated and programmable robot was invented by George Devol in 1954 and was ultimately called the Unimate. This ultimately laid the foundations of the modern robotics industry. Devol sold the first Unimate to General Motors in 1960, and it was installed in 1961 in a plant in Trenton, New Jersey to lift hot pieces of metal from a die casting machine and stack them. Devol's patent for the first digitally operated programmable robotic arm represents the foundation of the modern robotics industry.

The Rancho Arm was developed as a robotic arm to help handicapped patients at the Rancho Los Amigos Hospital in Downey, California; this computer controlled arm was bought by Stanford University in 1963. IBM announced its IBM System/360 in 1964. The system was heralded as being more powerful, faster, and more capable than its predecessors.

Marvin Minsky created the Tentacle Arm in 1968; the arm was computer controlled and its 12 joints were powered by hydraulics.Mechanical Engineering student Victor Scheinman created the Stanford Arm in 1969; the Stanford Arm is recognized as the first electronic computer controlled robotic arm (Unimate's instructions were stored on a magnetic drum).

The first mobile robot capable of reasoning about its surroundings, Shakey was built in 1970 by the Stanford Research Institute (now SRI International). Shakey combined multiple sensor inputs, including TV cameras, laser rangefinders, and bump sensors to navigate.

1970s

Freddy and Freddy II, both built in the United Kingdom, were robots capable of assembling wooden blocks in a period of several hours. German based company KUKA built the world's first industrial robot with six electromechanically driven axes, known as FAMULUS. In 1974, David Silver designed The Silver Arm; the Silver Arm was capable of fine movements replicating human hands. Feedback was provided by touch and pressure sensors and analyzed by a computer.

The SCARA, Selective Compliance Assembly Robot Arm, was created in 1978 as an efficient, 4-axis robotic arm. Best used for picking up parts and placing them in another location, the SCARA was introduced to assembly lines in 1981.

The Stanford Cart successfully crossed a room full of chairs in 1979. The Stanford Cart relied primarily on stereo vision to navigate and determine distances. The Robotics Institute at Carnegie Mellon University was founded in 1979 by Raj Reddy.

1980s

Takeo Kanade created the first "direct drive arm" in 1981. The first of its kind, the arm's motors were contained within the robot itself, eliminating long transmissions.

In 1984 Wabot-2 was revealed; capable of playing the organ, Wabot-2 had 10 fingers and two feet. Wabot-2 was able to read a score of music and accompany a person.

KUKA IR 160/60 Robots from 1983

In 1986, Honda began its humanoid research and development program to create robots capable of interacting successfully with humans. A hexapodal robot named Genghis was revealed by MIT in 1989. Genghis was famous for being made quickly and cheaply due to construction methods; Genghis used 4 microprocessors, 22 sensors, and 12 servo motors.Rodney Brooks and Anita M. Flynn published Fast, Cheap, and Out of Control: A Robot Invasion of The Solar System. The paper advocated creating smaller cheaper robots in greater numbers to increase production time and decrease the difficulty of launching robots into space.

1990s

The biomimetic robot RoboTuna was built by doctoral student David Barrett at the Massachusetts Institute of Technology in 1996 to study how fish swim in water. RoboTuna is designed to swim and resemble a blue fin tuna. Invented by Dr. John Adler, in 1994, the Cyberknife (a stereotactic radiosurgery performing robot) offered an alternative treatment of tumors with a comparable accuracy to surgery performed by human doctors.

IBM's Deep Blue computer, defeated World Chess Champion Garry Kasparov in 1997.

Honda's P2 humanoid robot was first shown in 1996. Standing for "Prototype Model 2", P2 was an integral part of Honda's humanoid development project; over 6 feet tall, P2 was smaller than its predecessors and appeared to be more human-like in its motions.

Expected to only operate for seven days, the Sojourner rover finally shuts down after 83 days of operation in 1997. This small robot (only weighing 23 lbs) performed semi-autonomous operations on the surface of Mars as part of the Mars Pathfinder mission; equipped with an obstacle avoidance program, Sojourner was capable of planning and navigating routes to study the surface of the planet. Sojourner's ability to navigate with little data about its environment and nearby surroundings allowed the robot to react to unplanned events and objects.

The P3 humanoid robot was revealed by Honda in 1998 as a part of the company's continuing humanoid project. In 1999, Sony introduced the AIBO, a robotic dog capable of interacting with humans, the first models released in Japan sold out in 20 minutes. Honda revealed the most advanced result of their humanoid project in 2000, named ASIMO. ASIMO is capable of running, walking, communication with humans, facial and environmental recognition, voice and posture recognition, and interacting with its environment. Sony also revealed its Sony Dream Robots, small humanoid robots in development for entertainment. In October 2000, the United Nations estimated that there were 742,500 industrial robots in the world, with more than half of the robots being used in Japan.

2001

Roomba vacuum cleaner docked in base station.

In April 2001, the Canadarm2 was launched an orbit and attached to the International Space Station. The Canadarm2 is a larger, more capable version of the arm used by the Space Shuttle and is hailed as being "smarter." Also in April, the Unmanned Aerial Vehicle Global Hawk made the first autonomous non-stop flight over the Pacific Ocean from Edwards Air Force Base in California to RAAF Base Edinburgh in Southern Australia. The flight was made in 22 hours.

The popular Roomba, a robotic vacuum cleaner, was first released in 2002 by the company iRobot.

In 2004, Cornell University revealed a robot capable of self-replication; a set of cubes capable of attaching and detaching, the first robot capable of building copies of itself. On 3 and 24 January the Mars rovers Spirit and Opportunity land on the surface of Mars. Launched in 2003, the two robots will drive many times the distance originally expected, and Opportunity is still operating as of mid 2012.

Self-driving cars had made their appearance by the middle of the first decade of the 21st century, but there was room for improvement. All 15 teams competing in the 2004 DARPA Grand Challenge failed to complete the course, with no robot successfully navigating more than five percent of the 150 mile off road course, leaving the $1 million prize unclaimed. In 2005, Honda revealed a new version of its ASIMO robot, updated with new behaviors and capabilities. In 2006, Cornell University revealed its "Starfish" robot, a 4-legged robot capable of self modeling and learning to walk after having been damaged. In 2007, TOMY launched the entertainment robot, i-sobot, which is a humanoid bipedal robot that can walk like a human beings and performs kicks and punches and also some entertaining tricks and special actions under "Special Action Mode".

Robonaut 2, the latest generation of the astronaut helpers, launched to the space station aboard Space Shuttle Discovery on the STS-133 mission. It is the first humanoid robot in space, and although its primary job for now is teaching engineers how dexterous robots behave in space, the hope is that through upgrades and advancements, it could one day venture outside the station to help spacewalkers make repairs or additions to the station or perform scientific work.

Commercial and industrial robots are now in widespread use performing jobs more cheaply or with greater accuracy and reliability than humans. They are also employed for jobs which are too dirty, dangerous or dull to be suitable for humans. Robots are widely used in manufacturing, assembly and packing, transport, earth and space exploration, surgery, weaponry, laboratory research, and mass production of consumer and industrial goods.

With recent advances in computer hardware and data management software, artificial representations of humans are also becoming widely spread. Examples include OpenMRS and EMRBots.

REFERENCES

1. Haug, Walter. "The Roman van Walewein as a postclassical literary experiment." In Originality and Tradition in the Middle Dutch Roman van Walewein, ed. B. Besamusca and E. Kooper. Cambridge, 1999. 17–28.

CHAPTER 2

Robot Kinematics

Chapter Contents

- **Introduction**
- **Kinematic Equations**
- **Forward Kinematics**
- **Kinematic Chain**
- **Inverse Kinematics**

Kinematics is the relationships between the positions, velocities, and accelerations of the links of a manipulator, where a manipulator is an arm, finger, or leg. In this chapter, we will develop a mathematical framework for describing these relationships, in particular position.

INTRODUCTION

Robot kinematics applies geometry to the study of the movement of multi-degree of freedom kinematic chains that form the structure of robotic systems. The emphasis on geometry means that the links of the robot are modeled as rigid bodies and its joints are assumed to provide pure rotation or translation.

Robot kinematics studies the relationship between the dimensions and connectivity of kinematic chains and the position, velocity and acceleration of each of the links in the robotic system, in order to plan and control movement and to compute actuator forces and torques. The relationship between mass and inertia properties, motion, and the associated forces and torques is studied as part of robot dynamics.

KINEMATIC EQUATIONS

A fundamental tool in robot kinematics is the kinematics equations of the kinematic chains that form the robot. These non-linear equations are used to map the joint parameters to the configuration of the robot system. Kinematics equations are also used in biomechanics of the skeleton and computer animation of articulated characters.

Forward kinematics uses the kinematic equations of a robot to compute the position of the end-effector from specified values for the joint parameters. The reverse process that computes the joint parameters that achieve a specified position of the end-effector is known as inverse kinematics. The dimensions of the robot and its kinematics equations define the volume of space reachable by the robot, known as its workspace.

There are two broad classes of robots and associated kinematics equations serial manipulators and parallel manipulators. Other types of systems with specialized kinematics equations are air, land, and submersible mobile robots, hyper-redundant, or snake, robots and humanoid robots.

FORWARD KINEMATICS

Forward kinematics refers to the use of the kinematic equations of a robot to compute the position of the end-effector from specified values for the joint parameters.

The kinematics equations of the robot are used in robotics, computer games, and animation. The reverse process that computes the joint parameters that achieve a specified position of the end-effector is known as inverse kinematics.

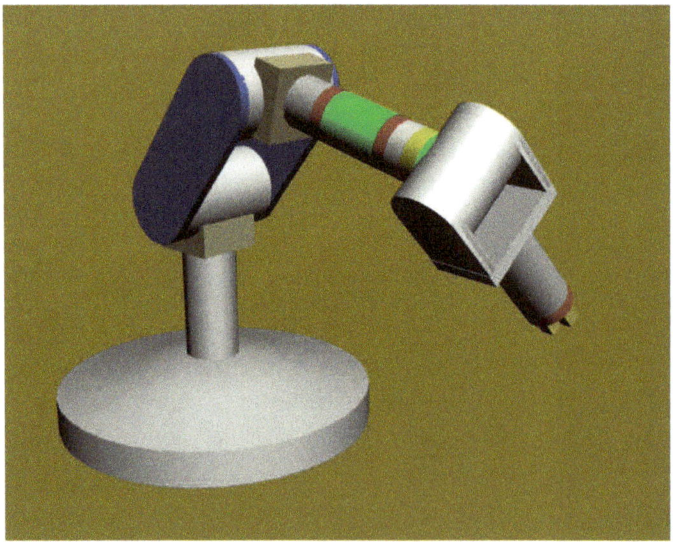

An articulated six DOF robotic arm uses forward kinematics to position the gripper.

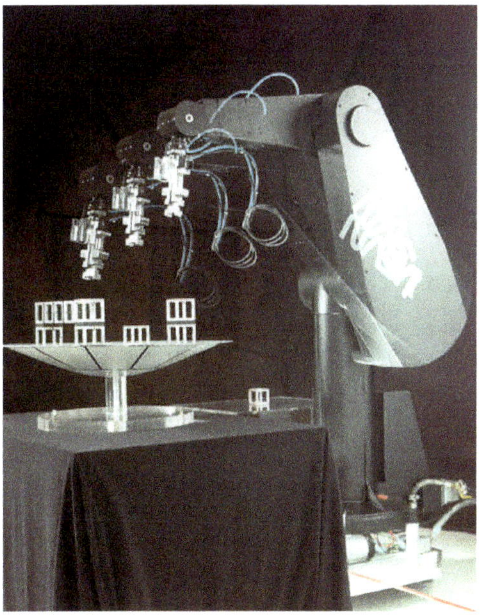

The forward kinematics equations define the trajectory of the end-effector of a PUMA robot reaching for parts.

Kinematics Equations

The kinematics equations for the series chain of a robot are obtained using a rigid transformation [Z] to characterize the relative movement allowed at each joint and separate rigid transformation [X] to define the dimensions of each link. The result is a sequence of rigid transformations alternating joint and link transformations from the base of the chain to its end link, which is equated to the specified position for the end link,

$[T] = [Z_1][X_1][Z_2][X_2]\ldots[X_{n-1}][Z_n]$, where T is the transformation locating the end-link. These equations are called the kinematics equations of the serial chain.

Link Transformations

In 1955, Jacques Denavit and Richard Hartenberg introduced a convention for the definition of the joint matrices [Z] and link matrices [X] to standardize the coordinate frame for spatial linkages. This convention positions the joint frame so that it consists of a screw displacement along the Z-axis

$$[Z_i] = \mathbf{Trans}_{Z_i}(d_i)\,\mathbf{Rot}_{Z_i}(\theta_i),$$

and it positions the link frame so it consists of a screw displacement along the X-axis,

$$[X_i] = \mathbf{Trans}_{X_i}(a_{i,i+1})\,\mathbf{Rot}_{X_i}(\alpha_{i,i+1}).$$

Using this notation, each transformation-link goes along a serial chain robot, and can be described by the coordinate transformation,

$$ {}^{i-1}T_i = [Z_i][X_i] = \mathbf{Trans}_{Z_i}(d_i)\,\mathbf{Rot}_{Z_i}(\theta_i)\,\mathbf{Trans}_{X_i}(a_{i,i+1})\,\mathbf{Rot}_{X_i}(\alpha_{i,i+1}), $$

where θ_i, d_i, $\alpha_{i,i+1}$ and $a_{i,i+1}$ are known as the Denavit-Hartenberg parameters.

Kinematics Equations Revisited

The kinematics equations of a serial chain of n links, with joint parameters θ_i are given by

$$ [T] = {}^0T_n = \prod_{i=1}^{n} {}^{i-1}T_i(\theta_i), $$

where ${}^{i-1}T_i(\theta_i)$ is the transformation matrix from the frame of link i to link i-1. In robotics, these are conventionally described by Denavit–Hartenberg parameters.

Denavit-Hartenberg Matrix

The matrices associated with these operations are:

$$ \operatorname{Trans}_{Z_i}(d_i) = \begin{bmatrix} 1 & 0 & 0 & 0 \\ 0 & 1 & 0 & 0 \\ 0 & 0 & 1 & d_i \\ 0 & 0 & 0 & 1 \end{bmatrix}, \quad \operatorname{Rot}_{Z_i}(\theta_i) = \begin{bmatrix} \cos\theta_i & -\sin\theta_i & 0 & 0 \\ \sin\theta_i & \cos\theta_i & 0 & 0 \\ 0 & 0 & 1 & 0 \\ 0 & 0 & 0 & 1 \end{bmatrix}. $$

Similarly,

$$ \operatorname{Trans}_{X_i}(a_{i,i+1}) = \begin{bmatrix} 1 & 0 & 0 & a_{i,i+1} \\ 0 & 1 & 0 & 0 \\ 0 & 0 & 1 & 0 \\ 0 & 0 & 0 & 1 \end{bmatrix}, \quad \operatorname{Rot}_{X_i}(\alpha_{i,i+1}) = \begin{bmatrix} 1 & 0 & 0 & 0 \\ 0 & \cos\alpha_{i,i+1} & -\sin\alpha_{i,i+1} & 0 \\ 0 & \sin\alpha_{i,i+1} & \cos\alpha_{i,i+1} & 0 \\ 0 & 0 & 0 & 1 \end{bmatrix}. $$

The use of the Denavit-Hartenberg convention yields the link transformation matrix, $[{}^{i-1}T_i]$ as

$$ {}^{i-1}T_i = \begin{bmatrix} \cos\theta_i & -\sin\theta_i\cos\alpha_{i,i+1} & \sin\theta_i\sin\alpha_{i,i+1} & a_{i,i+1}\cos\theta_i \\ \sin\theta_i & \cos\theta_i\cos\alpha_{i,i+1} & -\cos\theta_i\sin\alpha_{i,i+1} & a_{i,i+1}\sin\theta_i \\ 0 & \sin\alpha_{i,i+1} & \cos\alpha_{i,i+1} & d_i \\ 0 & 0 & 0 & 1 \end{bmatrix}, $$

known as the *Denavit-Hartenberg matrix.*

Computer Animation

The forward kinematic equations can be used as a method in 3D computer graphics for animating models.

The essential concept of forward kinematic animation is that the positions of particular parts of the model at a specified time are calculated from the position and orientation of the object, together with any information on the joints of an articulated model. So for example if the object to be animated is an arm with the shoulder remaining at a fixed location, the location of the tip of the thumb would be calculated from the angles of the shoulder, elbow, wrist, thumb and knuckle joints. Three of these joints (the shoulder, wrist and the base of the thumb) have more than one degree of freedom, all of which must be taken into account. If the model were an entire human figure, then the location of the shoulder would also have to be calculated from other properties of the model.

Forward kinematic animation can be distinguished from inverse kinematic animation by this means of calculation - in inverse kinematics the orientation of articulated parts is calculated from the desired position of certain points on the model. It is also distinguished from other animation systems by the fact that the motion of the model is defined directly by the animator - no account is taken of any physical laws that might be in effect on the model, such as gravity or collision with other models.

KINEMATIC CHAIN

Kinematic chain refers to an assembly of rigid bodies connected by joints to provide constrained (or desired) motion that is the mathematical model for a mechanical system. As in the familiar use of the word chain, the rigid bodies, or links, are constrained by their connections to other links. An example is the simple open chain formed by links connected in series, like the usual chain, which is the kinematic model for a typical robot manipulator.

Mathematical models of the connections, or joints, between two links are termed kinematic pairs. Kinematic pairs model the hinged and sliding joints fundamental to robotics, often called lower pairs and the surface contact joints critical to cams and gearing, called higher pairs. These joints are generally modeled as holonomic constraints. A kinematic diagram is a schematic of the mechanical system that shows the kinematic chain.

The modern use of kinematic chains includes compliance that arises from flexure joints in precision mechanisms, link compliance in compliant mechanisms and micro-electro-mechanical systems, and cable compliance in cable robotic and tensegrity systems.

The JPL mobile robot ATHLETE is a platform with six serial chain legs ending in wheels

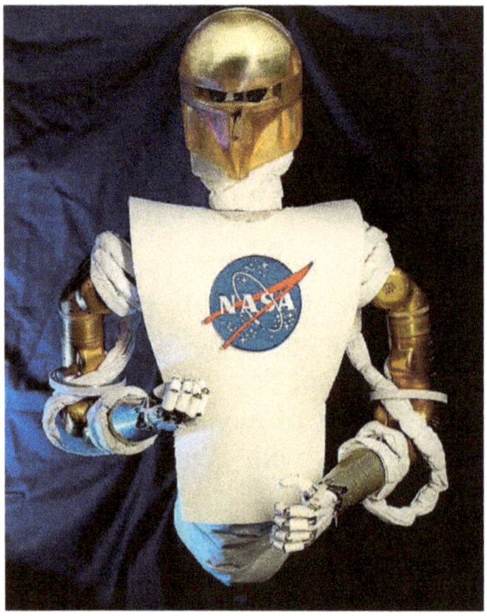

The arms, fingers and head of the JSC Robonaut are modeled as kinematic chains.

Mobility Formula

The degrees of freedom, or mobility, of a kinematic chain is the number of parameters that define the configuration of the chain. A system of n rigid bodies moving in space has 6n degrees of freedom measured relative to a fixed frame. This frame is included in the count of bodies, so that mobility does not depend on link that forms the fixed frame. This means the degree-of-freedom of this system is M=6(N-1), where N=n+1 is the number of moving bodies plus the fixed body.

Joints that connect bodies impose constraints. Specifically, hinges and sliders each impose five constraints and therefore remove five degrees of freedom. It is convenient to define the number of constraints c that a joint imposes in terms of the joint's freedom f, where c=6-f. In the case of a hinge or slider, which are one degree of freedom joints, have f=1 and therefore c=6-1=5.

The result is that the mobility of a kinematic chain formed from n moving links and j joints each with freedom f_i, i=1, ..., j, is given by

$$M = 6n - \sum_{i=1}^{j} (6 - f_i) = 6(N - 1 - j) + \sum_{i=1}^{j} f_i$$

Recall that N includes the fixed link.

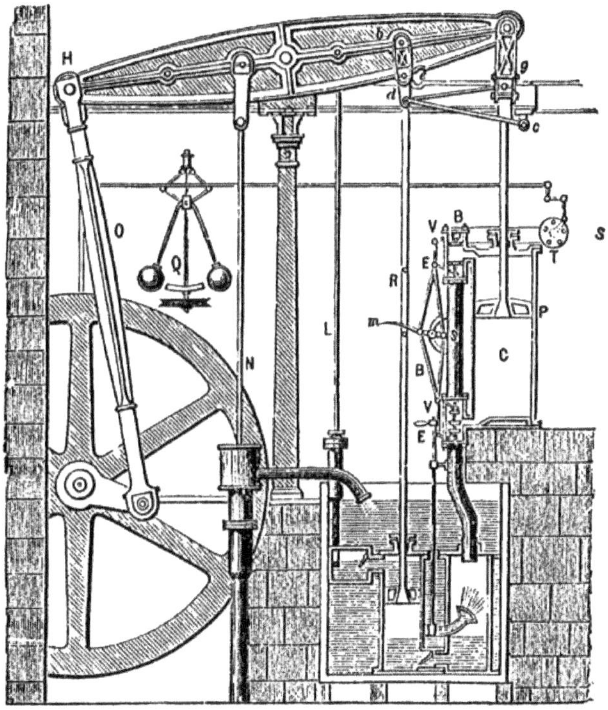

The movement of the Boulton & Watt steam engine is studied as a system of rigid bodies connected by joints forming a kinematic chain.

Analysis of Kinematic Chains

The constraint equations of a kinematic chain couple the range of movement allowed at each joint to the dimensions of the links in the chain, and form algebraic equations that are solved to determine the configuration of the chain associated with specific values of input parameters, called degrees of freedom.

The constraint equations for a kinematic chain are obtained using rigid transformations [Z] to characterize the relative movement allowed at each joint and separate rigid transformations [X] to define the dimensions of each link. In the case of a serial open chain, the result is a sequence of rigid transformations alternating joint and link transformations from the base of the chain to its end link, which is equated to the specified position for the end link. A chain of n links connected in series has the kinematic equations,

$$[T] = [Z_1][X_1][Z_2][X_2]\dots[X_{n-1}][Z_n],$$

where [T] is the transformation locating the end-link---notice that the chain includes a "zeroth" link consisting of the ground frame to which it is attached. These equations are called the forward kinematics equations of the serial chain.

Kinematic chains of a wide range of complexity are analyzed by equating the kinematics equations of serial chains that form loops within the kinematic chain. These equations are often called loop equations.

The complexity (in terms of calculating the forward and inverse kinematics) of the chain is determined by the following factors:

- Its topology: a serial chain, a parallel manipulator, a tree structure, or a graph.
- Its geometrical form: how are neighbouring joints spatially connected to each other?

Explanation:-Two or more rigid bodies in space are collectively called a rigid body system. We can hinder the motion of these independent rigid bodies with kinematic constraints. Kinematic constraints are constraints between rigid bodies that result in the decrease of the degrees of freedom of rigid body system.

Synthesis of Kinematic Chains

The constraint equations of a kinematic chain can be used in reverse to determine the dimensions of the links from a specification of the desired movement of the system. This is termed kinematic synthesis.

Perhaps the most developed formulation of kinematic synthesis is for four-bar linkages, which is known as Burmester theory.

Ferdinand Freudenstein is often called the father of modern kinematics for his contributions to the kinematic synthesis of linkages beginning in the 1950s. His use of the newly developed computer to solve Freudenstein›s equation became the prototype of computer-aided design systems.

This work has been generalized to the synthesis of spherical and spatial mechanisms.

INVERSE KINEMATICS

Inverse kinematics refers to the use of the kinematics equations of a robot to determine the joint parameters that provide a desired position of the end-effector. Specification of the movement of a robot so that its end-effector achieves a desired task is known as motion planning. Inverse kinematics transforms the motion plan into joint actuator trajectories for the robot.

The movement of a kinematic chain whether it is a robot or an animated character is modeled by the kinematics equations of the chain. These equations define the configuration of the chain in terms of its joint parameters. Forward kinematics uses the joint parameters to compute the configuration of the chain, and inverse kinematics reverses this calculation to determine the joint parameters that achieves a desired configuration.

For example, inverse kinematics formulas allow calculation of the joint parameters that position a robot arm to pick up a part. Similar formulas determine the positions of the skeleton of an animated character that is to move in a particular way.

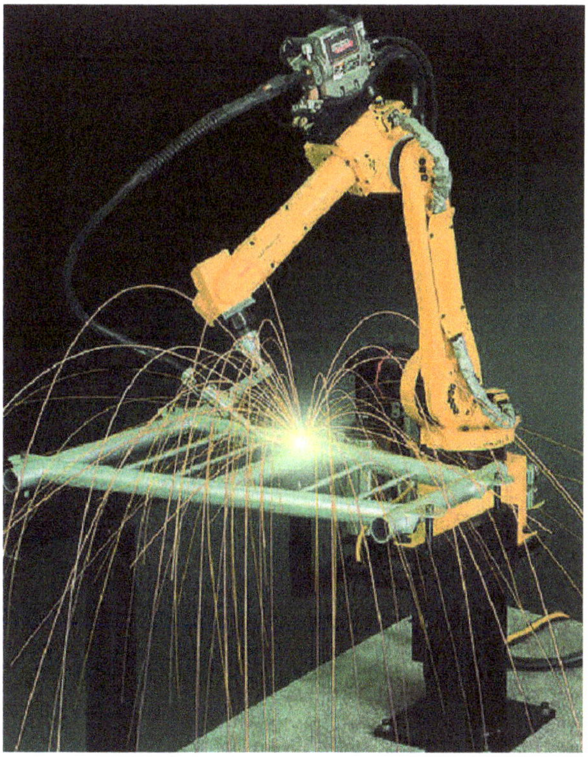

An industrial robot performing arc welding. Inverse kinematics computes the joint trajectories needed for the robot to guide the welding tip along the part.

Kinematic Analysis

Kinematic analysis is one of the first steps in the design of most industrial robots. Kinematic analysis allows the designer to obtain information on the position of each component within the mechanical system. This information is necessary for subsequent dynamic analysis along with control paths.

Inverse kinematics is an example of the kinematic analysis of a constrained system of rigid bodies, or kinematic chain. The kinematic equations of a robot can be used to define the loop equations of a complex articulated system. These loop equations are non-linear constraints on the configuration parameters of the system. The independent parameters in these equations are known as the degrees of freedom of the system.

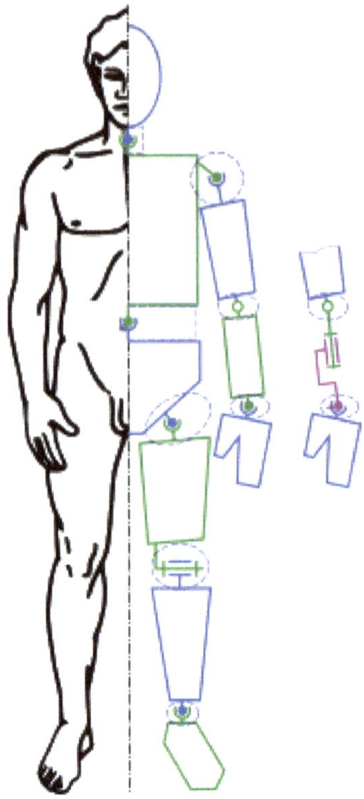

A model of the human skeleton as a kinematic chain allows positioning using inverse kinematics.

While analytical solutions to the inverse kinematics problem exist for a wide range of kinematic chains, computer modeling and animation tools often use Newton›s method to solve the non-linear kinematics equations.

Other applications of inverse kinematic algorithms include interactive manipulation, animation control and collision avoidance.

Inverse Kinematics and 3d Animation

Inverse kinematics is important to game programming and 3D animation, where it is used to connect game characters physically to the world, such as feet landing firmly on top of terrain.

An animated figure is modeled with a skeleton of rigid segments connected with joints, called a kinematic chain. The kinematics equations of the figure define the relationship between the joint angles of the figure and its pose or configuration. The forward kinematic animation problem uses the kinematics equations to determine the pose given the joint angles. The inverse kinematics problem computes the joint angles for a desired pose of the figure. It is often easier for computer-based designers, artists and animators to define the spatial configuration of an assembly or figure by moving parts, or arms and legs, rather than directly manipulating joint angles. Therefore, inverse kinematics is used in computer-aided design systems to animate assemblies and by computer-based artists and animators to position figures and characters.

The assembly is modeled as rigid links connected by joints that are defined as mates, or geometric constraints. Movement of one element requires the computation of the joint angles for the other elements to maintain the joint constraints. For example, inverse kinematics allows an artist to move the hand of a 3D human model to a desired position and orientation and have an algorithm select the proper angles of the wrist, elbow, and shoulder joints. Successful implementation of computer animation usually also requires that the figure move within reasonable anthropomorphic limits.

Analytical Solutions to Inverse Kinematics

An analytic solution to an inverse kinematics problem is a closed-form expression that takes the end-effector pose as input and gives joint positions as output, q = f(x). Analytical inverse kinematics solvers can be significantly faster than numerical solvers and provide more than one solution for a given end-effector pose.

The IKFast open-source program can solve for the complete analytical solutions of most common robot manipulators and generate C++ code for them. The generated solvers cover most degenerate cases and can finish in microseconds on recent computers.

Approximating Solutions To Ik Systems

There are many methods of modelling and solving inverse kinematics problems. The most flexible of these methods typically rely on iterative optimization to seek out an approximate solution, due to the difficulty of inverting the forward kinematics equation and the possibility of an empty solution space. The core idea behind several of these methods is to model the forward kinematics equation using a Taylor series expansion, which can be simpler to invert and solve than the original system.

The Jacobian Inverse Technique

The Jacobian inverse technique is a simple yet effective way of implementing inverse kinematics. Let there be m variables that govern the forward-kinematics equation, i.e. the position function. These variables may be joint angles, lengths, or other arbitrary real values. If the IK system lives in a 3-dimensional space, the position function can be viewed as a mapping $p(x): \Re^m \rightarrow \Re^3$. Let $p_0 = p(x_0)$ give the initial position of the system, and $p_1 = p(x_0 + \Delta x)$

be the goal position of the system. The Jacobian inverse technique iteratively computes an estimate of Δx that minimizes the error given by $||p(x_0 + \Delta x) - p_0||$.

For small Δx-vectors, the series expansion of the position function gives:

$$p(x_1) \approx p(x_0) + J_p(x_0)\Delta x$$

Where J_p (x0) is the (3 x m) Jacobian matrix of the position function at x_0.

Note that the (i, k)-th entry of the Jacobian matrix can be determined numerically:

$$\frac{\partial p_i}{\partial x_k} \approx \frac{p_i(x_{0,k}+h) - p_i(x_0)}{h}$$

Where $p_i(x)$ gives the i-th component of the position function, $x_{0,k}+h$ is simply x_0 with a small delta added to its k-th component, and h is a reasonably small positive value.

Taking the Moore-Penrose pseudoinverse of the Jacobian (computable using a singular value decomposition) and re-arranging terms results in:

$$\Delta x \approx J_p^+(x_0)\Delta p$$

Where $\Delta p = p(x_0 + \Delta x) - p(x_0)$.

Applying the inverse Jacobian method once will result in a very rough estimate of the desired Δx-vector. A line search should be used to scale this Δx to an acceptable value. The estimate for Δx can be improved via the following algorithm (known as the Newton-Raphson method):

$$\Delta x_{k+1} = J_p^+(x_k)\Delta p_k$$

Once some Δx-vector has caused the error to drop close to zero, the algorithm should terminate. Existing methods based on the Hessian matrix of the system have been reported to converge to desired Δx values using fewer iterations, though, in some cases more computational resources.

Robot Jacobian

The time derivative of the kinematics equations yields the Jacobian of the robot, which relates the joint rates to the linear and angular velocity of the end-effector. The principle of virtual work shows that the Jacobian also provides a relationship between joint torques and the resultant force and torque applied by the end-effector. Singular configurations of the robot are identified by studying its Jacobian.

Velocity Kinematics

The robot Jacobian results in a set of linear equations that relate the joint rates to the six-vector formed from the angular and linear velocity of the end-effector, known as a twist. Specifying the joint rates yields the end-effector twist directly.

The **inverse velocity** problem seeks the joint rates that provide a specified end-effector twist. This is solved by inverting the Jacobian matrix. It can happen that the robot is in a configuration where the Jacobian does not have an inverse. These are termed singular configurations of the robot.

Static Force Analysis

The principle of virtual work yields a set of linear equations that relate the resultant force-torque six vector, called a wrench, that acts on the end-effector to the joint torques of the robot. If the end-effector wrench is known, then a direct calculation yields the joint torques. The **inverse statics** problem seeks the end-effector wrench associated with a given set of joint torques, and requires the inverse of the Jacobian matrix. As in the case of inverse velocity analysis, at singular configurations this problem cannot be solved. However, near singularities small actuator torques result in a large end-effector wrench. Thus near singularity configurations robots have large mechanical advantage.

Fields of Study

Robot kinematics also deals with motion planning, singularity avoidance, redundancy, collision avoidance, as well as the kinematic synthesis of robots.

REFERENCES

1. Paul, Richard (1981). Robot manipulators: mathematics, programming, and control : the computer control of robot manipulators. MIT Press, Cambridge, MA. ISBN 978-0-262-16082-7.
2. J. M. McCarthy, 1990, Introduction to Theoretical Kinematics, MIT Press, Cambridge, MA.
3. John J. Craig, 2004, Introduction to Robotics: Mechanics and Control (3rd Edition), Prentice-Hall.
4. J. M. McCarthy and G. S. Soh, Geometric Design of Linkages, 2nd Edition, Springer 2010.
5. Paul, Richard (1981). Robot manipulators: mathematics, programming, and control : the computer control of robot manipulators. MIT Press, Cambridge, MA. ISBN 978-0-262-16082-7.
6. J. M. McCarthy, 1990, Introduction to Theoretical Kinematics, MIT Press, Cambridge, MA.
7. J. Denavit and R.S. Hartenberg, 1955, "A kinematic notation for lower-pair mechanisms based on matrices." Trans ASME J. Appl. Mech, 23:215–221.
8. Hartenberg, R. S., and J. Denavit. Kinematic Synthesis of Linkages. New York: McGraw-Hill, 1964 on-line through KMODDL
9. Jennifer Kay. "Introduction to Homogeneous Transformations & Robot Kinematics". Retrieved 2010-09-11.
10. Learn About Robots. "Robot Forward Kinematics". Retrieved 2007-02-01.
11. Paul, Richard (1981). Robot manipulators: mathematics, programming, and control : the computer control of robot manipulators. MIT Press, Cambridge, MA. ISBN 978-0-262-16082-7.
12. J. M. McCarthy, 1990, Introduction to Theoretical Kinematics, MIT Press, Cambridge, MA.
13. J. J. Uicker, G. R. Pennock, and J. E. Shigley, 2003, Theory of Machines and Mechanisms, Oxford University Press, New York.

14. J. M. McCarthy and G. S. Soh, 2010, Geometric Design of Linkages, Springer, New York.

CHAPTER 3

Robotic Sensing

Chapter Contents

Robotic sensing is the subarea of robotics science intended to give the robots sensing capabilities , so that the robots are more human-like , Robotic sensing mainly gives the robots the ability to see , touch , hear and move and it uses algorithms that require the environmental feedback .

INTRODUCTION

Robotic sensing is a subarea of robotics science intended to give robots sensing capabilities, so that robots are more human-like. Robotic sensing mainly gives robots the ability to see, touch, hear and move and uses algorithms that require environmental feedback.

Vision

Method

The visual sensing system can be based on anything from the traditional camera, sonar, and laser to the new technology radio frequency identification (RFID),which transmits radio signals to a tag on an object that emits back an identification code. All four methods aim for three procedures—sensation, estimation, and matching.

Image Processing

Image quality is important in applications that require excellent robotic vision. Algorithm based on wavelet transform for fusing images of different spectra and different foci improves image quality. Robots can gather more accurate information from the resulting improved image.

Usage

Visual sensors help robots to identify the surrounding and take appropriate action. Robots analyze the image of the immediate environment imported from the visual sensor. The result is compared to the ideal intermediate or end image, so that appropriate movement can be determined to reach the intermediate or final goal.

Touch

Signal Processing

Touch sensory signals can be generated by the robot's own movements. It is important to identify only the external tactile signals for accurate operations. Previous solutions employed the Wiener filter, which relies on the prior knowledge of signal statistics that are assumed to be stationary. Recent solution applies an adaptive filter to the robot's logic. It enables the robot to predict the resulting sensor signals of its internal motions, screening these false signals out. The new method improves contact detection and reduces false interpretation.

Usage

Touch patterns enable robots to interpret human emotions in interactive applications. Four measurable features—force, contact time, repetition, and contact area change—can effectively categorize touch patterns through the temporal decision tree classifier to account for the time delay and associate them to human emotions with up to 83% accuracy. The Consistency Index is applied at the end to evaluate the level of confidence of the system to prevent inconsistent reactions.

Robots use touch signals to map the profile of a surface in hostile environment such as a water pipe. Traditionally, a predetermined path was programmed into the robot. Currently, with the integration of touch sensors, the robots first acquire a random data point; the algorithm of the robot will then determine the ideal position of the next measurement according to a set of predefined geometric primitives. This improves the efficiency by 42%.

In recent years, using touch as a stimulus for interaction has been the subject of much study. In 2010, the robot seal PARO was built, which reacts to many stimuli from human interaction, including touch. The therapeutic benefits of such human-robot interaction is still being studied, but has shown very positive results.

Hearing

Signal Processing

Accurate audio sensor requires low internal noise contribution. Traditionally, audio sensors combine acoustical arrays and microphones to reduce internal noise level. Recent solutions combine also piezoelectric devices. These passive devices use the piezoelectric effect to transform force to voltage, so that the vibration that is causing the internal noise could be eliminated. On average, internal noise up to about 7dB can be reduced.

Robots may interpret strayed noise as speech instructions. Current voice activity detection (VAD) system uses the complex spectrum circle centroid (CSCC) method and a maximum signal-to-noise ratio (SNR) beamformer. Because humans usually look at their partners when conducting conversations, the VAD system with two microphones enable the robot to locate the instructional speech by comparing the signal strengths of the two microphones. Current system is able to cope with background noise generated by televisions and sounding devices that come from the sides.

Usage

Robots can perceive our emotion through the way we talk. Acoustic and linguistic features are generally used to characterize emotions. The combination of seven acoustic features and four linguistic features improves the recognition performance when compared to using only one set of features.

Acoustic Feature

- Duration
- Energy
- Pitch
- Spectrum
- Cepstral
- Voice quality
- Wavelets

Linguistic feature

- Bag of words
- Part-of-speech

- Higher semantics
- Varia

Movement

Usage

Automated robots require a guidance system to determine the ideal path to perform its task. However, in the molecular scale, nano-robots lack such guidance system because individual molecules cannot store complex motions and programs. Therefore, the only way to achieve motion in such environment is to replace sensors with chemical reactions. Currently, a molecular spider that has one streptavidin molecule as an inert body and three catalytic legs is able to start, follow, turn and stop when came across different DNA origami. The DNA-based nano-robots can move over 100 nm with a speed of 3 nm/min.

In a TSI operation, which is an effective way to identify tumors and potentially cancer by measuring the distributed pressure at the sensor's contacting surface, excessive force may inflict a damage and have the chance of destroying the tissue. The application of robotic control to determine the ideal path of operation can reduce the maximum forces by 35% and gain a 50% increase in accuracy compared to human doctors.

Performance

Efficient robotic exploration saves time and resources. The efficiency is measured by optimality and competitiveness. Optimal boundary exploration is possible only when a robot has square sensing area, starts at the boundary, and uses the Manhattan metric. In complicated geometries and settings, a square sensing area is more efficient and can achieve better competitiveness regardless of the metric and of the starting point.

SENSOR-BASED SORTING

Sensor-based sorting, is an umbrella term for all applications where particles are singularly detected by a sensor technique and rejected by an amplified mechanical, hydraulic or pneumatic process. The technique is generally applied in the three industries mining, recycling and food processing and used in the particle size range between 0.5 and 300 mm (0.020 and 11.811 in). Since sensor-based sorting is a single particle separation technology, the throughput is proportional to the average particle size and weight fed onto the machine.

Parallel installation of two belt-type sensor-based ore sorters at Mittersill mine

Functional Principle of Sensor-Based Sorting

The main subprocesses of sensor-based sorting are material conditioning, material presentation, detection, data processing and separation. Material conditioning includes all operations which prepare the particles for being detected by the sensor. All optical sensors need clean material to be able to detect optical characteristics. Conditioning includes screening and cleaning of the feed material. The aim of the material presentation is the isolation of the particles by creating a single particle layer with the densest surface cover possible without particles touching each other and enough distance to each other allowing for a selective detection and rejection of each single particle. There are two types of sensor-based sorters: the chute type and the belt type. For both types the first step in acceleration is spreading out the particles by a vibrating feeder followed by either a fast belt or a chute. On the belt type the sensor usually detects the particles horizontally while they pass it on the belt. For the chute type the material detection is usually done vertically while the material passes the sensor in a free fall. The data processing is done in real time by a computer. The computer transfers the result of the data processing to an ultra fast ejection unit which, depending on the sorting decision, ejects a particle or lets it pass.

Sensor-Based Ore Sorting

Sensor-based ore sorting is the terminology used for sensor-based sorting in the mining industry. It is a coarse physical coarse particle separation technology usually applied in the size range for 25–100 mm (0.98–3.94 in). Aim is either to create a lumpy product in ferrous metals, coal or industrial minerals applications or to reject waste before it enters production bottlenecks and more expensive comminution and concentration steps in the process. In the majority of all mining processes, particles of sub-economic grade enter the traditional comminution, classification and concentration steps. If the amount of sub-economic material in the above-mentioned fraction is roughly 25% or more, there is good potential that sensor-based ore sorting is a technically and financially viable option. High added value can be achieved with relatively low capital expenditure, especially when increasing the productivity through downstream processing of higher grade feed and through increased overal recovery when rejecting deleterious waste.

Introduction

Sensor-based sorting is a coarse particle separation technology applied in mining for the dry separation of bulk materials. The functional principle does not limit the technology to any kind of segment or mineral application but makes the technical viability mainly depend on the liberation characteristics at the size range 25–100 mm (0.98–3.94 in), which is usually sorted. If physical liberation is present there is a good potential that one of the sensors available on industrial scale sorting machines can differentiate between valuable and non-valuable particles.

The separation is based on features measured with a detection technology that are used to derive a yes/no decision for actuation of usually pneumatic impulses. Sensor-based sorting is a disruptive technology in the mining industry which is universally applicable for all commodities.

A comprehensive study examines both the technology's potential and its limitations, whilst providing a framework for application development and evaluation. All relevant aspects, from sampling to plant design and integration into mining and mineral processing systems, are covered. Other terminologeis used in the industry include *ore sorting, automated sorting, electronic sorting,* and *optical sorting*.

Definition

Sensor-based sorting has been introduced by Wotruba and Harbeck as an umbrella term for all applications where particles are singularly detected by a sensor technique and then rejected by an amplified mechanical, hydraulic or pneumatic process.

Liberation Characteristics

As for any other physical separation process, liberation is pre-requisite for possible separation. Liberation characteristics are well known and relatively easy to study for particulate lots in smaller size ranges, e.g. flotation feed and products. The analysis is essential for understanding the possible results of physical separation and relatively easy to conduct in laboratory on a couple of dozens of grams of sample which can be studied using optical methods or such as the QEMSCAN. For larger particles above 10 mm (0.39 in) it is widely known for applications that are treated using density separation methods, such as coal or iron ore.

Here, the washability analysis can be conducted on sample masses up to 10 tonnes in equipped laboratories. For sensor-based sorting, where laboratory methods can only tell about the liberation characteristics where the describing feature is the density (e.g. iron ore, coal), hand counting, single-particle tests and bulk tests can reveal the liberation characteristics of a bulk material: Hereby, only single particle tests reveal the true liberation, while hand counting and bulk testing give a result which also incorporates the separation efficiency of the type of analysis. More information on the testing procedures used in technical feasibility evaluation can be found in the respective chapter.

Historical Development

The oldest form of mineral processing practiced since the Stone Age is hand-picking. Georgius Agricola also describes hand-picking is his book *De re metallica* in 1556. Sensor-based sorting is the automation and extension to hand picking. In addition to sensors that measure visible differences like color and the further interpretation of the data regarding texture and shape, other sensors are available on industrial scale sorters that are able to measure differences invisible for the human eye (EM, XRT, NIR). The principles of the technology and the first machinery has been developed since the 1920s (. Nevertheless, widely applied and standard technology it is only in the industrial minerals and gemstone segments. Mining is benefiting from the step change developments in sensing and computing technologies and from machine development in the recycling and food processing industries. In 2002, Cutmore and Eberhard stated that the relatively small installed base of sensor-based sorters in mining is more a result of insufficient industry interest than any technical barriers to their effective use Nowadays sensor-based sorting is beginning to reveal its potential in various applications in basically all segments of mineral production (industrial minerals, gemstones, base-metals, precious metals, ferrous metals, fuel).

Precondition is physical liberation in coarse size ranges (~10–300 mm (0.39–11.81 in)) to make physical separation possible. Either the product fraction, but more often the waste fraction needs to be liberated. If liberation is present, there is good potential that one of available detection technologies on today›s sensor-based sorters can positively or negatively identify one of the two desired fractions.

Feed Preparation

A size range coefficient of approximately three is advisable. A minimum amount of undersized fine material must enter the machines to optimize availability. Moisture of the feed is not important, if the material is sufficiently dewatered and the undersize fraction is efficiently removed. For surface detection technologies sometimes spray water on the classifying screen is required to clean the surfaces. Surface detection technologies would otherwise measure the reflectance of the adhesions on the surface and a correlation to the particle's content is not given.

Main Machine Types

During the more than 80 years of technical development of sensor-based ore sorting equipment, various types of machines have been developed. This includes the channel-type, bucket-wheel type and cone type sorters. The main machine types being installed in the mining industry today are belt-type and chute-type machines. Harbeck made a good comparison of both disadvantages and advantages of the systems for different sorting applications. The selection of a machine-type for an application depends various case-dependent factors, including the detection system applied, particle size, moisture, yield amongst others.

Chute-Type Machine

The chute-type machine has a lower footprint and fewer moving parts which results in lower investment and operating costs. In general, it is more applicable to well liberated material and surface detection, because a double sided scanning is possible on a more reliable on the system. The applicable top size of the chute-type machine is bigger, as material handling of particles up to 300 mm (12 in) is only technically viable on this setup.

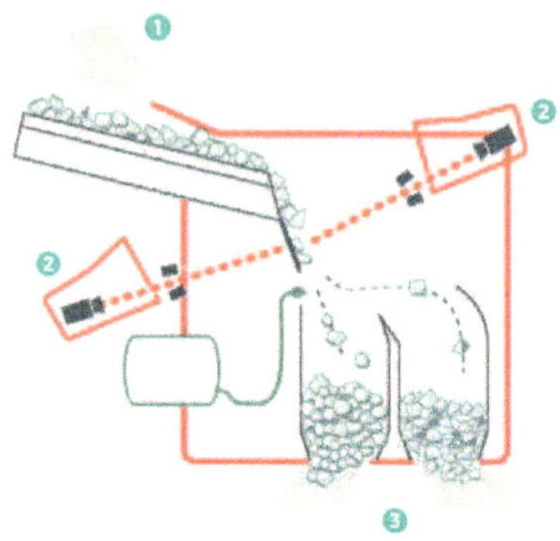

Side view and working principle of chute-type sensor-based ore sorting machine

Belt-Type Machine

The belt-type machine is generally more applicable to smaller and to adhesive feed. In addition, the feed presentation is more stable which makes it more applicable for more difficult and heterogienous applications.

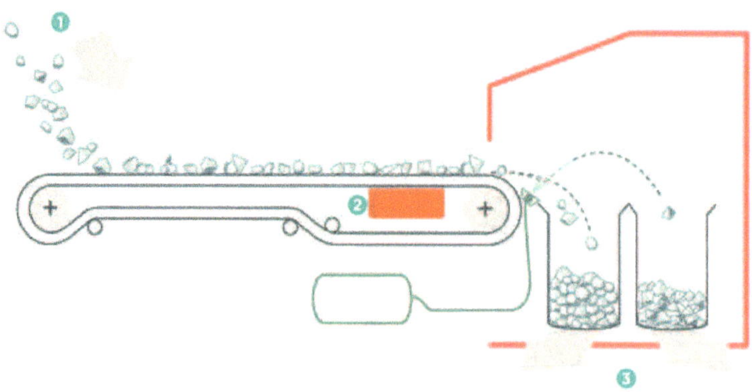

Side view and working principle of belt-type sensor-based ore sorting machine

Sub-Processes

The separation in both machine types comprises the following sub-processes:

Feed Distribution

A sized screen fraction with a size range coefficient (d95/d5) of 2-5 (optimal 2-3) is fed onto a vibratory feeder which has the function to create a mono-layer, by pre-accelerating the particles. A common misunderstanding in plant design is, that you can use the vibratory feeder to discharge from a buffer bunker but a separate units needs to be applied, since the feed distribution is very important to the efficiency of the sensor-based sorter and different loads on the feeder change its position and vibration characteristics.

Presentation

The feed is then transferred to the presentation mechanism which is the belt or the chute in the two main machine types respectively. This sub-process has the function to pass single particles of the material stream in a stable and predictable manner, thus in a unidirectional movement orthogonal to the detection line with uniform speed profile.

Detection

In the detection sub-process location and property vectors are recorded to allow particle localization for ejection and material classification for discrimination purposes. All detection technologies applied have in common to be cheap, contactless and fast. The technologies are subdivided in transmitting and reflecting groups, the first measuring the inner content of a particle while the later only uses the surface reflection for discrimination. Surface, or reflection technologies have the disadvantage that the surfaces need to be representing the content, thus need to be clean from clay and dust adhesions. But by default surface reflection technologies violate the Fundamental Sampling Principle because not all components of a particle have the same probability of being detected. The main transmitting technologies are EM (Electromagnetics) and XRT (X-ray-Transmission). EM detection is based on the conductivity of the material passing an alternating electromagnetic field. The principle of XRT is widely known through the application in medical diagnostics and airport luggage scanners. The main surface or reflection technologies are traditionally X-ray luminescence detectors capturing the fluorescence of diamonds under the excitation of X-ray radiation and color cameras detecting brightness and colour difference. Spectroscopic methods such as near-infrared spectroscopy known from remote sensing in exploration in mining for decades, have found their way into industrial scale sensor-based sorters. Advantage of the application of near-infrared spectroscopy is that the evidence can be measured on the presence of specific molecular bonds, thus minerals composition of the near-infrared active minerals. There is more detection technologies available on industrial scale sensor-based ore sorters. Readers that want to go into detail can find more in the literature.

Data Processing

Spectral and spatial is collected by the detection system. The spatial component catches the position of the particles distribution across the width of the sorting machine, which is then used in case the ejection mechanism is activated for a single particle. Spectral data comprises the features that are used for material discrimination. In a superseding processing step, spectral and spatial can be combined to include patterns into the separation criterion. Huge amount of data is collected in real time multiple processing and filtering steps are bringing the data down to the Yes/no decision – either for ejecting a particle or for keeping the ejection mechanism still for that one.

Ejection

The state-of-the-art mechanism of today's sensor-based ore sorters is a pneumatic ejection. Here, a combination of high speed air valves and an array of nozzles perpendicular to the acceleration belt or chute allows precise application of air pulses to change the direction of flight of single particles. The nozzle pitch and diameter is adapted to the particle size. The air impulse must be precise enough to change the direction of flight of a single particle by applying the drag force to this single particle and directing it over the mechanical splitter plate.

Installations

Sensor-based sorting installations normally comprise the following basic units; crusher, screen, sensor-based sorter and compressor. There are principally two different kinds of installations that are described in the following paragraphs – stationary and semi-mobile installations.

Semi-Mobile Installations

Transportable semi-mobile installations have gained increasing popularity in the last two decades. They are enabled by the fact that complete sensor-based sorting systems are relatively compact in relation to the capacity in tonnes per hour. This is mainly because little infrastructure is needed. The picture shows a containerised sensor-based sorter which is applied in Chromitite sorting. The system is operated in conjunction with a Diesel-powered mobile crusher and screen. Material handling of the feed, undersize fraction, product and waste fraction is conducted using a wheel loader. The system is powered by a Diesel generator and a compressor station delivers the instrument quality air needed for the operation. Semi-mobile installations are applied primarily to minimise material handling and save transport costs. Another reason for choosing the semi-mobile option for an installation is bulk testing of new ore bodies. Capacity of a system very much depends on the size fraction sorted, but a 250tph capacity is a good estimate for semi-mobile installations, considering a capacity of 125tph sorter feed and 125tph undersize material. During the last decade both generic plant designs and customised designs have been developed, for example in the framework of the i2mine project.

Containerised sensor-based ore sorter applied in a modular plant operated as semi-mobile installation

Stationary Installations

To cope with high volume mass flows and for application, where a changing physical location of the sensor-based sorting process is of no benefit for the financial feasibility of the operation, stationary installations are applied. Another reason for applying stationary installations are multistage (Rougher, Scavenger, Cleaner) sensor-based ore sorting processes. Within stationary installations, sorters are usually located in parallel, which allows transport of the discharge fractions with one product and one waste belt respectively, which decreases plant footprint and amount of conveyors.

Positions in the Flow-Sheet

Lumpy ore Production

For higher grade applications such as ferrous metals, coal and industrial minerals, sensor-based ore sorting can be applied to create a final product. Pre-condition is, that the liberation allows for the creation of a sellable product. Undersize material is usually bypassed as product, but can also be diverted to the waste fraction, if the composition does not meet the required specifications. This is case and application dependant.

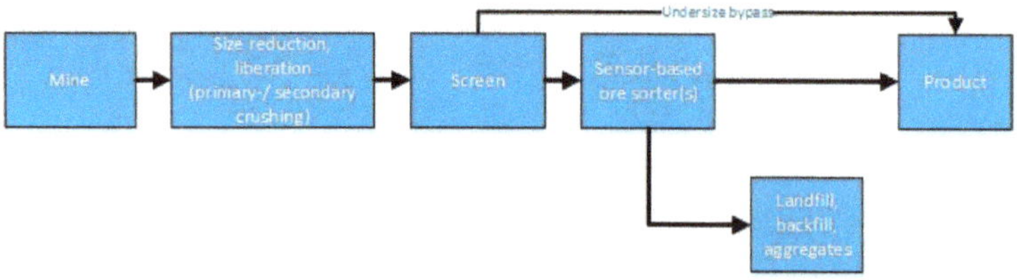

Simplified flow-sheet for lumpy ore production

Waste Rejection

Most prominent example of the application of sensor-based ore sorting is the rejection of barren waste before transporting and comminution. Waste rejection is also known under the term pre-concentration. A discrimination has been introduced by Robben. Rule of thumb is that at least 25% of liberated barren waste must be present in the fraction to be treated by sensor-based ore sorting to make waste rejection financially feasible. Reduction of waste before it enters comminution and grinding processes does not only reduce the costs in those processes, but also releases the capacity that can be filled with higher grade material and thus implies higher productivity of the system. A prejudice against the application of a waste rejection process is, that the valuable content lost in this process is a penalty higher than the savings that can be achieved. But it is reported in the literature that the overall recovery even increases through bringing higher grade material as feed into the mill. In addition, the higher productivity is an additional source of income. If noxious waste such as acid consuming calcite is removed, the downstream recovery increases and the downstream costs decrease disproportionally as reported for example by Bergmann. The coarse waste rejected can be an additional source of income if there is a local market for aggregates.

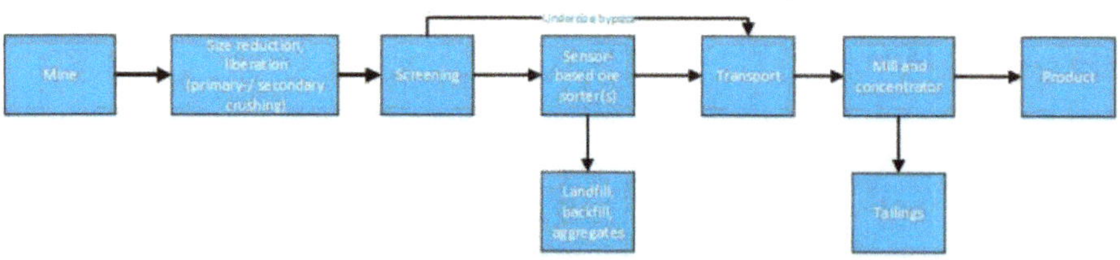

Simplified flow-sheet for waste rejection with sensor-based ore sorting

Marginal Ore Treatment

Sensor-based ore sorting is financially especially attractive for low grade or marginal ore or waste dump material. This described scenario describes that waste dump material or marginal ore is sorted and added to the run-of-mine production. The needed capacity for the sensor-based ore sorting step is less in this case such as the costs involved. Requirement is that two crude material streams are fed in parallel, requiring two crushing stations. Alternatively, marginal and high grade ore can be buffered on an intermediate stockpile and dispatched in an alternating operation. The latter option has the disadvantage that the planned production time, the loading, of the sensor-based ore sorter is low, unless a significant intermediate stockpile or bunker is installed. Treating the marginal ore separately has the advantage that less equipment is needed since the processed material stream is lower, but is has the disadvantage that the potential of the technology is not unfolded for the higher grade material, where sensor-based sorting would also add benefit.

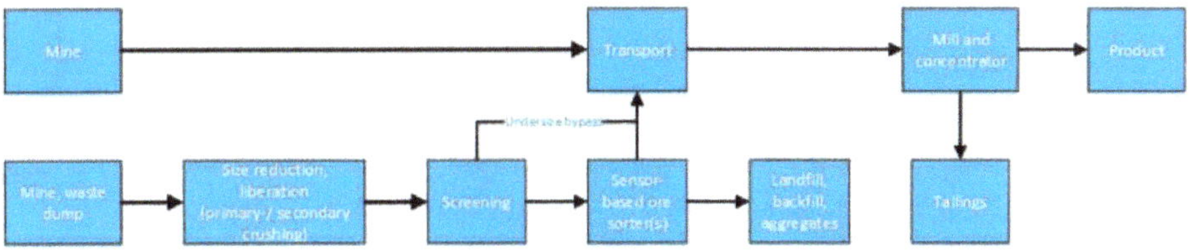

Simplified flow-sheet for marginal ore treatment with sensor-based ore sorting

Sensor-Based Ore Sorting in Pebble-Circuit of Autogenious and Semi-Autogenious Mills

Pebble circuits are a very advandageous location for the application of sensor-based ore sorters. Usually it is hard waste recirculating and limiting the total mill capacity. In addition, the tonnage is significantly lower in comparison to the total run-of-mine stream, the size range is applicable and usually uniform and the particles' surfaces are clean. High impact on total mill capacity is reported in the literature.

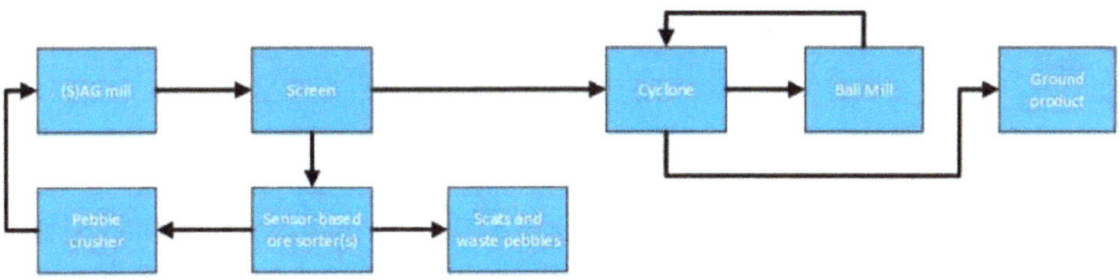

Simplified flow-sheet for waste and scats removal with sensor-based ore sorting from a pebble circuit

Ore-Type Diversion

Sensor-based sorting can be applied to separate the coarse fraction of the run-of-mine material according to its characteristics. Possible separation criteria are grade, mineralogy, grade and grindability amongst others. Treating different ore types separately results either in an optimised cash flow in the sense, that revenue ist shifted to an earlier point in time, or increased overall recovery which translates to higher productivity and thus revenue. If two separate plant lines are installed, the increased productivity must compensate for the overall higher capital expenditure and operating costs.

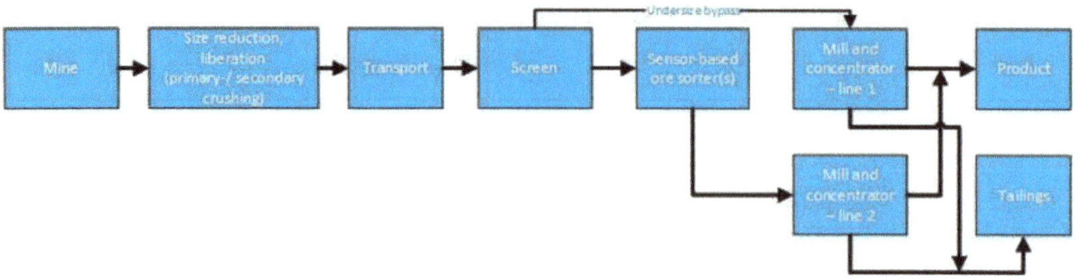

Simplified flow-sheet of ore type diversion with sensor-based ore sorting

Economic Impact

Sensor-based ore sorting is in comparison to other coarse particle separation technologies relatively cheap. While the costs for the equipment itself are relatively high in capital expenditure and operating costs, the absence of extensive infrastructure in a system results in operating costs that are to be compared to jigging. The specific costs are very much depending on the average particle size of the feed and on the ease of the separation. Coarser particles imply higher capacity and thus less costs. Detailed costing can be conducted after the mini-bulk stage in the technical feasibility evaluation. Prejudice against waste rejection with sensor-based sorting widely spread, that the loss of valuables, thus the recovery penalty of this process, supersedes the potential downstream cost savings and is therefore economically not viable. It must be noted that for waste rejection the aim for the separation with sensor-based ore sorting must be put onto maximum recovery, which means that only low grade or barren waste is rejected because the financial feasibility is very much sensitive to that factor. Nevertheless, through the rejection of waste before comminution and concentration steps, recovery can be often increased in the downstream process, meaning that the overall recovery is equal or even higher than the one in the base case, meaning that instead of losing product, additional product can be produced, which adds the additional revenue to the cost savings on the positive side in the cash flow. If the rejected material is replaced with additional higher grade material, the main economic benefit unfolds through the additional production. It implies, that in conjunction with sensor-based ore sorting, the capacity of the crushing station is increased, to allow for the additional mass-flow that is subsequently taken out by the sensor-based ore sorters as waste.

Technical Feasibility Testing

Bulk Characterisation

Precondition for the applicability of sensor-based ore sorting is the presence of liberation at the particle size of interest. Before entering into sensor-based ore sorting testing procedures there is the possibility to assess the degree of liberation through the inspection of drill cores, hand-counting and washability analysis. The quantification of liberation does not include any process efficiencies, but gives an estimate of the possible sorting result and can thus be applied for desktop financial feasibility analysis. Drill core analysis Both for green-field and brown-field applications, inspection of drill core in combination with the grade distribution and mineralogical description is a good option for estimation of the liberation characteristics and the possible success of sensor-based ore sorting. In combination with the mining method and mine plan, an estimation of possible grade distribution in coarse particles can be done.

Hand Counting

Hand-counting is a cheap and easy to conduct method to estimate the liberation characteristics of a bulk sample wither originating from run-of-mine material, a waste dump or for example exploration trenching. Analysis of particles in the size range 10-100mm has been conducted on a total sample mass of 10 tonnes. By visual inspection of trained personnel, a classification of each particle into different bins (e.g. lithology, grade) is possible and the distribution is determined by weighing each bin. A trained professional can quickly estimate the efficiency of a specific detection and process efficiency of sensor-based ore sorting knowing the sensor response of the mineralogy of ore in question and other process efficiency parameters.

Washability Analysis

The washability analysis is widely known in bulk material analysis, where the specific density is the physical property describing the liberation and the separation results, which is then in the form of the partition curve. The partition curve is defined as the curve which gives as a function of a physical property or characteristic, the proportions in which different elemental classes of raw feed having the same property are split into separate products. It is thus per its definition not limited to, but predominantly applied in analysis of liberation and process efficiency of density separation processes. For sensor-based ore sorting, the partition (also called Tromp) curves for chromite, iron ore and coal are known and can thus be applied for process modelling.

Single Particle Testing

Single particle testing is an extensive but powerful laboratory procedure developed by Tomra. Outo of a sample set of multiple hundreds of fragments in the size range 30-60mm are measured individually on each of the available detection technologies. After recording of the raw data, all the fragments are comminute and assayed individually which then allows plotting of the liberation function of the sample set and in addition, the detection efficiency of each detection technology in combination with the calibration method applied. This makes the evaluation of detection and calibration and subsequently the selection of the most powerful combination possible. This analysis is possible to be applied on quarters or half sections of drill core.

(Mini-) Bulk Testing

Mini-bulk tests are conducted with 1-100t of samples on industrial scale sensor-based ore sorters. The size fraction intervals to be treated are prepared using screen classifications. Full capacity is established then with each fraction and multiple cut-points are programmed in the sorting software. After creating multiple sorting fractions in rougher, scavenger and cleaner steps these weighed are sent for assays. The resulting data delivers all input for flow-sheet development. Since the tests are conducted on industrial scale equipment, there is no scale-up factor involved when designing a flow-sheet and installation of sensor-based ore sorting.

Bulk/ Pilot Testing

To gather relevant statistical data higher sample masses are needed in some cases. Thus, transportation of the sample into the mini-bulk testing facility becomes unviable and the equipment is set up in the field. Containerised units in conjunction with Diesel-powered crushing and screening equipment are often applied and used for production test runs under full scale operating conditions.

Process Efficiency

The process efficiency of sensor-based ore sorting is described in detail by C. Robben in 2014. The total process efficiency is subdivided into the following sub-process efficiencies; Platform efficiency, preparation efficiency, presentation efficiency, detection efficiency and separation efficiency. All the sub-process contribute to the total process efficiency, of course in combination with the liberation characteristics of the bulk material that the technology is applied to. The detailed description of the sib-processes and their contribution to the total process efficiency can be found in the literature.

Suppliers

Steinert GmbH – Sorting technologies for Mining and Recycling

With more than 125 years of expertise in primary and secondary raw material sorting, Steinert is the oldest company in the world which provides sorting technologies and innovative solutions for different markets, like recycling and mining. Steinert has developed a large variety of sensors, like X-ray, inductive, NIR and color optical sensors and 3D laser camera, which can be combined for sorting even the most difficult materials. For the recycling field, the HSI technology – Hyper Spectral Imaging –, is the state-of-the-art in NIR sorting technology.

Tomra Sorting Solutions | Mining

A sensor-based sorting equipment supplier with large installed base in the industries mining, recycling and food. Tomra's sensor-based sorting equipment and services for the precious metals and base metals segment are marketed through a cooperation agreement with Outotec from Finland, which brings the extensive comminution, processing and application experience of Outotec together with Tomra's sensor-based ore sorting technology and application expertise.

Sensor-Based Sorting Conference

The expert conference "Sensor-Based Sorting" is addressing new developments and applications in the field of automatic sensor separation techniques for primary and secondary raw materials. The conference provides a platform for plant operators, manufacturers, developers and scientists to exchange know-how and experiences. The congress is hosted by the Department of Processing and Recycling and the Unit for Mineral Processing (AMR) of RWTH Aachen University in cooperation with the GDMB Society of Metallurgists and Miners, Clausthal. Scientific supervisors are Professor Thomas Pretz and Professor Hermann Wotruba. Researchers and practitioners in the field of sensor-based sorting are invited to submit an abstract. Authors of accepted abstracts will be asked to submit full papers for the conference proceedings. All topics are for oral presentation at the conference. Topics may include theoretical and practical aspects, methodological and empirical approaches as well as case studies and research activities in sensor-based sorting of minerals and waste. Application-oriented contributions are welcome.

Videos

Tungsten plays a large and indispensable role in modern high-tech industry. Up to 500,000 tons of raw tungsten ore are mined each year by Wolfram Bergbau und Hütten AG (WHB) in Felbertal, Austria, which is the largest scheelite deposit in Europe. 25% of the run-of-mine ore are separated as waste before entering the mill.

REFERENCES

1. Roh SG, Choi HR (Jan 2009). "3-D Tag-Based RFID System for Recognition of Object." IEEE Transactions on Automation Science and Engineering 6 (1): 55–65.
2. Arivazhagan S, Ganesan L, Kumar TGS (Jun 2009). "A modified statistical approach for image fusion using wavelet transform." Signal Image and Video Processing 3 (2): 137-144.
3. Jafar FA, et al (Mar 2011). "An Environmental Visual Features Based Navigation Method for Autonomous Mobile Robots." International Journal of Innovative

Computing, Information and Control 7 (3): 1341-1355.

4. Anderson S, et al (Dec 2010). "Adaptive Cancelation of Self-Generated Sensory Signals in a Whisking Robot." IEEE Transactions on Robotics 26 (6): 1065-1076.
5. Kim YM, et al (Aug 2010)."A Robust Online Touch Pattern Recognition for Dynamic Human-robot Interaction." IEEE Transactions on Consumer Electronics 56 (3): 1979-1987.
6. Mazzini F, et al (Feb 2011). "Tactile Robotic Mapping of Unknown Surfaces, with Application to Oil Wells." IEEE Transactions on Instrumentation and Measurement 60 (2): 420-429.
7. Matsumoto M, Hashimoto S (2010). "Internal Noise Reduction Using Piezoelectric Device under Blind Condition." Internatl (Jan 2011). "Searching for the most important feature types signalling emotion-related user states in speech." Computer Speech and Language 25 (1): 4-28.
8. Lund K, et al (May 2010). "Molecular robots guided by prescriptive landscapes." Nature 465 (7295): 206-210.
9. Trejos AL, et al (Sep 2009). "Robot-assisted Tactile Sensing for Minimally Invasive Tumor Localization." International Journal of Robotics Research 28 (9): 1118-1133.
10. Czyzowicz J, Labourel A, Pelc A (Jan 2011). "Optimality and Competitiveness of Exploring Polygons by Mobile Robots." Information and Computation 209 (1): 74-88.
11. https://www.youtube.com/watch?v=oJq5PQZHU-I
12. Kim HD, et al (2009). "Target Speech Detection and Separation for Communication with Humanoid Robots in Noisy Home Environments." Advanced Robotics 23 (15): 2093-2111.
13. Batliner A, et al (Jan 2011). "Searching for the most important feature types signalling emotion-related user states in speech." Computer Speech and Language 25 (1): 4-28.
14. Syntouch technology
15. . Tactile sensors also come in the form of pressure indicating films that reveal pressure distribution and magnitude between contacting surfaces by virtue of an immediate and permanent color change. These pressure indicating films are one-time use sensor that capture the maximum pressure they were exposed to. Pressure indicating films are activated by chemical reaction and are non-electronic sensors. Robotic Tactile Sensing - Technologies and System
16. Tactile Sensing - From Humans to Humanoids
17. A tactile sensor sheet using pressure conductive rubber with electrical-wires stitched method
18. Towards Tactile Sensing System on Chip for Robotic Applications
19. Piezoelectric oxide semiconductor field effect transistor touch sensing devices
20. Data sheet for Schunk FT-Nano 43, a 6-axis force torque sensor

21. https://www.researchgate.net/profile/Mathew_Evans2/publication/221116457_Tactile_Discrimination_Using_Template_Classifiers_Towards_a_Model_of_Feature_Extraction_in_Mammalian_Vibrissal_Systems/links/0deec516be39482f27000000.pdf
22. A robust micro-vibration sensor for biomimetic fingertips
23. Development of a tactile sensor based on biologically inspired edge encoding
24. A biologically inspired tactile sensor array utilizing phase-based computation
25. TakkTile sensor plans
26. DIY tactile sensor based on webcam
27. Kleine, Christopher; Wotruba, Hermann and Robben, Mathilde: A new tool for mining engineers – The sensor-based sorter. Sustainable Development in the Minerals Industry (SDIMI) Conference 2011, Aachen, Germany, ISBN 978-3-86797-119-5
28. Wotruba, Hermann: Sensor Sorting Technology – Is the Minerals Industry Missing a Chance?. Proceedings of XXIII International Minerals Processing Congress, IMPC 2006, Istanbul, Turkey
29. Kleine, Christopher and Wotruba, Hermann: Added value to the mining industry by the integration of sensor based sorting. Aachen international mining symposia, Mineral Resources and Mine Development, Institute of Mining Engineering I, May 26th and 27th 2010, 411-434
30. Robben C, Characteristics of sensor-based sorting technology and implementation in mining . - Shaker Verlag GmbH, in der Schriftenreihe zur Aufbereitung und Veredlung, ISBN 978-3-8440-2498-2
31. Wotruba, Hermann, und Hartmut Harbeck. "Sensor-Based Sorting." In Ullmann's Encyclopedia of Industrial Chemistry. Wiley-VCH Verlag GmbH & Co. KGaA, 2010.
32. Agricola G De Re Metallica Libri XII . - Wiebaden : Fourier Verlag GmbH 2003, 1556.
33. Sweet A T Patent: 1678884. - United States of America, 1928.
34. Cutmore, N. G. and Eberhardt, J. E. "The future of ore sorting in sustainable processing". Cairns, Australia : s.n., 2002. Processing Conference 2002.
35. Clayton C G and Spackman R Gold ore sorting Nr. 4830193 Patent - United States of America, Dec 29 1987.
36. Blagden T and Verboomen J Clean coal at the speed of light Report - Caboolture, Australia : Australian Coal Industry's Research Program (ACARP) Project No C13052, 2009.
37. Harbeck H and Kroog H New developments in sensor-based sorting.
38. Wotruba, H., Robben, M.R., Balthasar, D., Near-infrared sensor-based sorting in the minerals industry. In Proc. Conference in minerals engineering. Luleå Technical University, Lulea (Sweden), 2009, pp. 163–176)
39. i2mine Innovative technologies and concepts for the intelligent deep mine of

the future – European Commission under the Cooperation Programme of the 7th Framework Programme for Research and Technological Development in the 'Nanosciences, Nanotechnologies, Materi Online // Project overview. - 2012. - 22 May 2012. - http://www.i2mine.eu/.

40. Bergmann, J. Sensor-Based Sorting – Experience, Technology and Opportunities. Falmouth, UK : Minerals Engineering International, 2011. Proceedings of Physical Separation 2011.
41. K. Seerane (Rio Tinto Group) and G. Rech (CommodasUltrasort), "Investigation of sorting technology to remove hard pebbles and recover copper bearing rocks from an autogenious mill circuit", The Southern African Institute of Mining and Metallurgy (SAIMM) 6th Southern African Base Metals Conference 2011, Phalaborwa, Republic of South Africa
42. Sanders G J The principles of coal preparation . - Newcastle, Australia: Australian Coal Preparation Society, 2007. - Vol. IV.
43. Department of Processing and Recycling. "Sensor-Based Sorting 2014". RWTH Aachen University. Retrieved 4 March 2016.
44. Tungsten Sorting at WOLFRAM Bergbau AG, Austria. YouTube. 13 May 2014. Retrieved 4 March 2016 – via YouTube.

CHAPTER 4

Degrees of Freedom Mechanics

Chapter Contents

Degrees of freedom (mechanics). In physics, the degree of freedom (DOF) of a mechanical system is the number of independent parameters that define its configuration.

INTRODUCTION

In physics, the degree of freedom (DOF) of a mechanical system is the number of independent parameters that define its configuration. It is the number of parameters that determine the state of a physical system and is important to the analysis of systems of bodies in mechanical engineering, aeronautical engineering, robotics, and structural engineering.

The position of a single car (engine) moving along a track has one degree of freedom because the position of the car is defined by the distance along the track. A train of rigid cars connected by hinges to an engine still has only one degree of freedom because the positions of the cars behind the engine are constrained by the shape of the track.

An automobile with highly stiff suspension can be considered to be a rigid body traveling on a plane (a flat, two-dimensional space). This body has three independent degrees of freedom consisting of two components of translation and one angle of rotation. Skidding or drifting is a good example of an automobile›s three independent degrees of freedom.

The position and orientation of a rigid body in space is defined by three components of translation and three components of rotation, which means that it has six degrees of freedom.

The exact constraint mechanical design method manages the degrees of freedom to neither underconstrain nor overconstrain a device.

MOTIONS AND DIMENSIONS

The position of an n-dimensional rigid body is defined by the rigid transformation, $[T] = [A, d]$, where d is an n-dimensional translation and A is an $n \times n$ rotation matrix, which has n translational degrees of freedom and $n(n - 1)/2$ rotational degrees of freedom. The number of rotational degrees of freedom comes from the dimension of the rotation group SO(n).

A non-rigid or deformable body may be thought of as a collection of many minute particles (infinite number of DOFs), this is often approximated by a finite DOF system. When motion involving large displacements is the main objective of study (e.g. for analyzing the motion of satellites), a deformable body may be approximated as a rigid body (or even a particle) in order to simplify the analysis.

The degree of freedom of a system can be viewed as the minimum number of coordinates required to specify a configuration. Applying this definition, we have:

- For a single particle in a plane two coordinates define its location so it has two degrees of freedom;
- A single particle in space requires three coordinates so it has three degrees of freedom;
- Two particles in space have a combined six degrees of freedom;
- If two particles in space are constrained to maintain a constant distance from each other, such as in the case of a diatomic molecule, then the six coordinates must satisfy a single constraint equation defined by the distance formula. This reduces the degree of freedom of the system to five, because the distance formula can be used to solve for the remaining coordinate once the other five are specified.

SIX DEGREES OF FREEDOM

Six degrees of freedom (6DoF) refers to the freedom of movement of a rigid body in three-dimensional space. Specifically, the body is free to change position as forward/backward (surge), up/down (heave), left/right (sway) translation in three perpendicular axes, combined with changes in orientation through rotation about three perpendicular axes, often termed pitch, yaw, and roll.

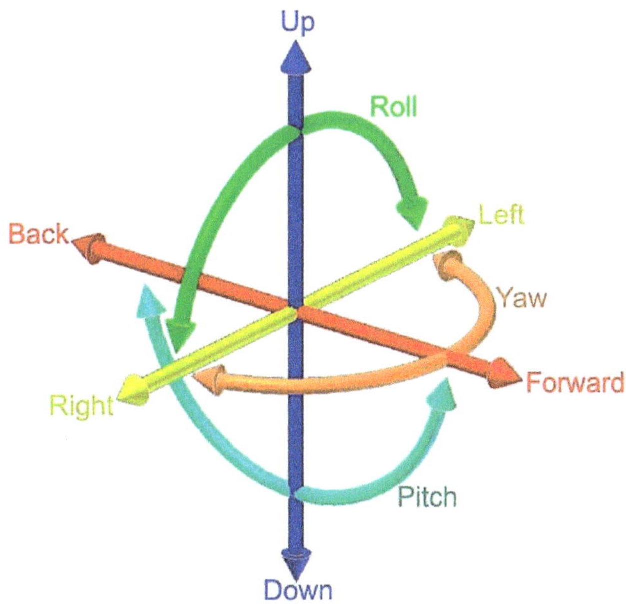

The six degrees of freedom: forward/back, up/down, left/right, pitch, yaw, roll

Robotics

Serial and parallel manipulator systems are generally designed to position an end-effector with six degrees of freedom, consisting of three in translation and three in orientation. This provides a direct relationship between actuator positions and the configuration of the manipulator defined by its forward and inverse kinematics.

Robot arms are described by their degrees of freedom. This number typically refers to the number of single-axis rotational joints in the arm, where higher number indicates an increased flexibility in positioning a tool. This is a practical metric, in contrast to the abstract definition of degrees of freedom which measures the aggregate positioning capability of a system.

In 2007, Dean Kamen, inventor of the Segway, unveiled a prototype robotic arm with 14 degrees of freedom for DARPA. Humanoid robots typically have 30 or more degrees of freedom, with six degrees of freedom per arm, five or six in each leg, and several more in torso and neck.

Engineering

The term is important in mechanical systems, especially biomechanical systems for analyzing and measuring properties of these types of systems that need to account for all six degrees of freedom. Measurement of the six degrees of freedom is accomplished today through both AC and DC magnetic or electromagnetic fields in sensors that transmit positional and angular data to a processing unit. The data are made relevant through software that integrate the data based on the needs and programming of the users. Ascension Technology Corporation has recently created a 6DoF device small enough to fit in a biopsy needle, allowing physicians to better research at minute levels. The new sensor passively senses pulsed DC magnetic fields generated by either a cubic transmitter or a flat transmitter and is available for integration and manufacturability by medical OEMs.

An example of six degree of freedom movement is the motion of a ship at sea. It is described as :

Translation:

- Moving forward and backward on the X-axis. (Surging)
- Moving left and right on the Y-axis. (Swaying)
- Moving up and down on the Z-axis. (Heaving)

Rotation

- Tilting side to side on the X-axis. (Rolling)
- Tilting forward and backward on the Y-axis. (Pitching)
- Turning left and right on the Z-axis. (Yawing)

Operational Envelope Types

There are three types of operational envelope in the Six degrees of freedom. These types are *Direct*, *Semi-direct* (conditional) and *Non-direct,* all regardless of the time remaining for the execution of the maneuver, the energy remaining to execute the maneuver and finally, if the motion is commanded via a biological entity (human) or a robotical entity (computer).

1 *Direct type* : Involved a degree can be commanded directly without particularly conditions and described as a normal operation. (An aileron on a basic airplane)

2 *Semi-direct* type : Involved a degree can be commanded when some specific conditions are met. (Reverse thrust on an aircraft)

3 *Non-direct* type : Involved a degree when is achieved via the interaction with its environment and cannot be commanded. (Pitching motion of a vessel at sea)

Transitional type also exists in some vehicles. For example, when the Space Shuttle operates in space, the craft is described as fully-direct-six because its six degrees can be commanded. However, when the Space Shuttle is in the earth›s atmosphere for its return, the fully-direct-six degrees are not longer applicable for many technical reasons.

Game Controllers

First-person shooter (FPS) games generally provide five degrees of freedom: forwards/ backwards, slide left/right, up/down (jump/crouch/lie), yaw (turn left/right), and pitch (look up/down). If the game allows leaning control, then some consider it a sixth DoF; however, this may not be completely accurate, as a lean is a limited partial rotation.

The term *6DoF* has sometimes been used to describe games which allow freedom of movement, but do not necessarily meet the full 6DoF criteria. For example, *Dead Space 2*, and to a lesser extent, *Homeworld* and *Zone Of The Enders* allow freedom of movement.

Some examples of true 6DoF games, which allow independent control of all three movement axes and all three rotational axes, include *Shattered Horizon*, the *Descent* franchise, *Retrovirus*, *Miner Wars*, *Space Engineers*, *Forsaken* and *Overload* (from the same creators of *Descent*). The space MMO *Vendetta Online* also features 6 degrees of freedom.

Motion tracking devices such as TrackIR are used for 6DoF head tracking. This device often finds its places in flight simulators and other vehicle simulators that require looking around the cockpit to locate enemies or simply avoiding accidents in-game.

The acronym **3DoF**, meaning movement in the three dimensions but not rotation, is sometimes encountered.

The Razer Hydra, a motion controller for PC, tracks position and rotation of two wired nunchucks, providing six degrees of freedom on each hand.

The SpaceOrb 360 is a 6DOF computer input device released in 1996 originally manufactured and sold by the SpaceTec IMC company (first bought by Labtec, which itself was later bought by Logitech).

The motion of a ship at sea has the six degrees of freedom of a rigid body, and is described as:

Translation and rotation:

- Moving up and down (elevating/heaving);
- Moving left and right (strafing/swaying);
- Moving forward and backward (walking/surging);
- Swivels left and right (yawing) ;
- Tilts forward and backward (pitching);
- Pivots side to side (rolling).

The trajectory of an airplane in flight has three degrees of freedom and its attitude along the trajectory has three degrees of freedom, for a total of six degrees of freedom.

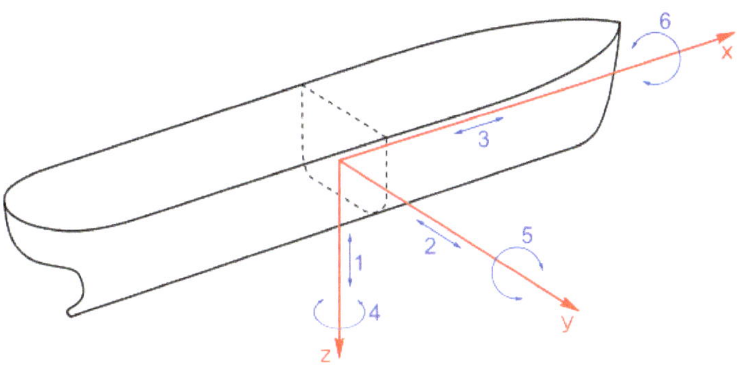

The six degrees of freedom of movement of a ship

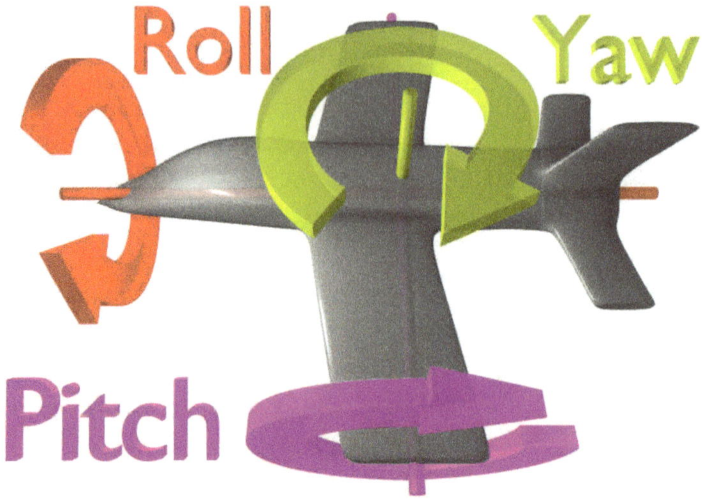

Attitude degrees of freedom for an airplane

Mobility Formula

The mobility formula counts the number of parameters that define the configuration of a set of rigid bodies that are constrained by joints connecting these bodies.

Consider a system of n rigid bodies moving in space has $6n$ degrees of freedom measured relative to a fixed frame. In order to count the degrees of freedom of this system, include the ground frame in the count of bodies, so that mobility is independent of the choice of the body that forms the fixed frame. Then the degree-of-freedom of the unconstrained system of $N = n + 1$ is

$$M = 6n = 6(N-1),$$

because the fixed body has zero degrees of freedom relative to itself.

Joints that connect bodies in this system remove degrees of freedom and reduce mobility. Specifically, hinges and sliders each impose five constraints and therefore remove five degrees of freedom. It is convenient to define the number of constraints c that a joint imposes in terms of the joint's freedom f, where $c = 6 - f$. In the case of a hinge or slider, which are one degree of freedom joints, have $f = 1$ and therefore $c = 6 - 1 = 5$.

The result is that the mobility of a system formed from n moving links and j joints each with freedom f_i, $i = 1, \ldots, j$, is given by

$$M = 6n - \sum_{i=1}^{j} (6 - f_i) = 6(N - 1 - j) + \sum_{i=1}^{j} f_i$$

Recall that N includes the fixed link.

There are two important special cases: (i) a simple open chain, and (ii) a simple closed chain. A single open chain consists of n moving links connected end to end by n joints, with one end connected to a ground link. Thus, in this case $N = j + 1$ and the mobility of the chain is

$$M = \sum_{i=1}^{j} f_i$$

For a simple closed chain, n moving links are connected end-to-end by n + 1 joints such that the two ends are connected to the ground link forming a loop. In this case, we have N = j and the mobility of the chain is

$$M = \sum_{i=1}^{j} f_i - 6$$

An example of a simple open chain is a serial robot manipulator. These robotic systems are constructed from a series of links connected by six one degree-of-freedom revolute or prismatic joints, so the system has six degrees of freedom.

An example of a simple closed chain is the RSSR spatial four-bar linkage. The sum of the freedom of these joints is eight, so the mobility of the linkage is two, where one of the degrees of freedom is the rotation of the coupler around the line joining the two S joints.

Planar and Spherical Movement

It is common practice to design the linkage system so that the movement of all of the bodies are constrained to lie on parallel planes, to form what is known as a *planar linkage*. It is also possible to construct the linkage system so that all of the bodies move on concentric spheres, forming a *spherical linkage*. In both cases, the degrees of freedom of the links in each system is now three rather than six, and the constraints imposed by joints are now $c = 3 - f$.

In this case, the mobility formula is given by

$$M = 3(N - 1 - j) + \sum_{i=1}^{j} f_i,$$

and the special cases become

- planar or spherical simple open chain,

$$M = \sum_{i=1}^{j} f_i,$$

- planar or spherical simple closed chain,

$$M = \sum_{i=1}^{j} f_i - 3.$$

An example of a planar simple closed chain is the planar four-bar linkage, which is a four-bar loop with four one degree-of-freedom joints and therefore has mobility $M = 1$.

Systems of Bodies

A system with several bodies would have a combined DOF that is the sum of the DOFs of the bodies, less the internal constraints they may have on relative motion. A mechanism or linkage containing a number of connected rigid bodies may have more than the degrees of freedom for a single rigid body. Here the term *degrees of freedom* is used to describe the number of parameters needed to specify the spatial pose of a linkage. A specific type of linkage is the open kinematic chain, where a set of rigid links are connected at joints; a joint may provide one DOF (hinge/sliding), or two (cylindrical). Such chains occur commonly in robotics, biomechanics, and for satellites and other space structures. A human arm is considered to have seven DOFs. A shoulder gives pitch, yaw, and roll, an elbow allows for pitch, and a wrist allows for pitch,yaw and roll . Only 3 of those movements would be necessary to move the hand to any point in space, but people would lack the ability to grasp things from different angles or directions. A robot (or object) that has mechanisms to control all 6 physical DOF is said to be holonomic. An object with fewer controllable DOFs than total DOFs is said to be non-holonomic, and an object with more controllable DOFs than total DOFs (such as the human arm) is said to be redundant. Although keep in mind that it is not redundant in the human arm because the two DOFs; wrist and shoulder, that represent the same movement; roll, supply each other since they can›t do a full 360. The degree of freedom are like different movements that can be made.

In mobile robotics, a car-like robot can reach any position and orientation in 2-D space, so it needs 3 DOFs to describe its pose, but at any point, you can move it only by a forward motion and a steering angle. So it has two control DOFs and three representational DOFs; i.e. it is non-holonomic. A fixed-wing aircraft, with 3–4 control DOFs (forward motion, roll, pitch, and to a limited extent, yaw) in a 3-D space, is also non-holonomic, as it cannot move directly up/down or left/right.

A summary of formulas and methods for computing the degrees-of-freedom in mechanical systems has been given by Pennestri, Cavacece, and Vita.

Electrical Engineering

In electrical engineering *degrees of freedom* is often used to describe the number of directions in which a phased array antenna can form either beams or nulls. It is equal to one less than the number of elements contained in the array, as one element is used as a reference against which either constructive or destructive interference may be applied using each of the remaining antenna elements. Radar practice and communication link practice, with beam steering being more prevalent for radar applications and null steering being more prevalent for interference suppression in communication links.

REFERENCES

1. http://ocw.mit.edu/courses/mechanical-engineering/2-76-multi-scale-system-design-fall-2004/readings/reading_l3.pdf

2. Summary of ship movement Archived November 25, 2011, at the Wayback Machine.
3. J. J. Uicker, G. R. Pennock, and J. E. Shigley, 2003, Theory of Machines and Mechanisms, Oxford University Press, New York.
4. J. M. McCarthy and G. S. Soh, Geometric Design of Linkages, 2nd Edition, Springer 2010
5. Pennestri E, Cavacece M, Vita L, On the computation of degrees-of-freedom: A didactic perspective, ASME Paper DETC2005-84109
6. Paul, Richard P., Robot Manipulators: Mathematics, Programming, and Control, MIT Press, 1981.
7. Craig, John J., Introduction to Robotics: Mechanics and Control, Addison-Wesley, 1986.
8. Medical News Today. Ascension Develops World's Smallest Six Degrees-of-Freedom Sensor For Emerging Medical Procedures. 25 Aug 2008
9. Summary of ship movement Archived November 25, 2011, at the Wayback Machine.

CHAPTER 5

Three Laws of Robotics

Chapter Contents

A robot may not injure a human being or, through inaction, allow a human being to come to harm. A robot must obey orders given it by human beings except where such orders would conflict with the First Law. A robot must protect its own existence as long as such protection does not conflict with the First or Second Law.

INTRODUCTION

The Three Laws of Robotics (often shortened to The Three Laws or known as Asimov's Laws) are a set of rules devised by the science fiction author Isaac Asimov. The rules were introduced in his 1942 short story "Runaround", although they had been foreshadowed in a few earlier stories. The Three Laws, quoted as being from the "Handbook of Robotics, 56th Edition, 2058 A.D.", are:

1. A robot may not injure a human being or, through inaction, allow a human being to come to harm.

2. A robot must obey the orders given it by human beings except where such orders would conflict with the First Law.

3. A robot must protect its own existence as long as such protection does not conflict with the First or Second Laws.

These form an organizing principle and unifying theme for Asimov's robotic-based fiction, appearing in his Robot series, the stories linked to it, and his Lucky Starr series of young-adult fiction. The Laws are incorporated into almost all of the positronic robots appearing in his fiction, and cannot be bypassed, being intended as a safety feature. Many of Asimov's robot-focused stories involve robots behaving in unusual and counter-intuitive ways as an unintended consequence of how the robot applies the Three Laws to the situation in which it finds itself. Other authors working in Asimov's fictional universe have adopted them and references, often parodic, appear throughout science fiction as well as in other genres.

This cover of I, Robot illustrates the story "Runaround", the first to list all Three Laws of Robotics.

The original laws have been altered and elaborated on by Asimov and other authors. Asimov himself made slight modifications to the first three in various books and short stories to further develop how robots would interact with humans and each other. In later fiction where robots had taken responsibility for government of whole planets and human civilizations, Asimov also added a fourth, or zeroth law, to precede the others:

0. A robot may not harm humanity, or, by inaction, allow humanity to come to harm.

The Three Laws, and the zeroth, have pervaded science fiction and are referred to in many books, films, and other media.

In The Rest of the Robots, published in 1964, Asimov noted that when he began writing in 1940 he felt that "one of the stock plots of science fiction was ... robots were created and destroyed their creator. Knowledge has its dangers, yes, but is the response to be a retreat from knowledge? Or is knowledge to be used as itself a barrier to the dangers it brings?" He decided that in his stories robots would not "turn stupidly on his creator for no purpose but to demonstrate, for one more weary time, the crime and punishment of Faust."

On May 3, 1939, Asimov attended a meeting of the Queens (New York) Science Fiction Society where he met Ernest and Otto Binder who had recently published a short story "I, Robot" featuring a sympathetic robot named Adam Link who was misunderstood and motivated by love and honor. (This was the first of a series of ten stories; the next year "Adam Link's Vengeance" (1940) featured Adam thinking "A robot must never kill a human, of his own free will.") Asimov admired the story. Three days later Asimov began writing "my own story of a sympathetic and noble robot", his 14th story. Thirteen days later he took "Robbie" to John W. Campbell the editor of Astounding Science-Fiction. Campbell rejected it claiming that it bore too strong a resemblance to Lester del Rey's "Helen O'Loy", published in December 1938; the story of a robot that is so much like a person that she falls in love with her creator and becomes his ideal wife. Frederik Pohl published "Robbie" in Astonishing Stories magazine the following year.

Asimov attributes the Three Laws to John W. Campbell, from a conversation that took place on 23 December 1940. Campbell claimed that Asimov had the Three Laws already in his mind and that they simply needed to be stated explicitly. Several years later Asimov's friend Randall Garrett attributed the Laws to a symbiotic partnership between the two men – a suggestion that Asimov adopted enthusiastically. According to his autobiographical writings Asimov included the First Law's "inaction" clause because of Arthur Hugh Clough's poem "The Latest Decalogue", which includes the satirical lines "Thou shalt not kill, but needst not strive / officiously to keep alive".

Although Asimov pins the creation of the Three Laws on one particular date, their appearance in his literature happened over a period. He wrote two robot stories with no explicit mention of the Laws, "Robbie" and "Reason". He assumed, however, that robots would have certain inherent safeguards. "Liar!", his third robot story, makes the first mention of the First Law but not the other two. All three laws finally appeared together in "Runaround". When these stories and several others were compiled in the anthology I, Robot, "Reason" and "Robbie" were updated to acknowledge all the Three Laws, though the material Asimov added to "Reason" is not entirely consistent with the Three Laws as he described them elsewhere. In particular the idea of a robot protecting human lives when it does not believe those humans truly exist is at odds with Elijah Baley's reasoning, as described below.

During the 1950s Asimov wrote a series of science fiction novels expressly intended for young-adult audiences. Originally his publisher expected that the novels could be adapted into a long-running television series, something like The Lone Ranger had been for radio. Fearing that his stories would be adapted into the "uniformly awful" programming he saw flooding the television channels Asimov decided to publish the Lucky Starr books under the pseudonym "Paul French". When plans for the television series fell through, Asimov decided to abandon the pretence; he brought the Three Laws into Lucky Starr and the Moons of Jupiter, noting that this "was a dead giveaway to Paul French's identity for even the most casual reader".

In his short story "Evidence" Asimov lets his recurring character Dr. Susan Calvin expound a moral basis behind the Three Laws. Calvin points out that human beings are typically expected to refrain from harming other human beings (except in times of extreme duress like war, or to save a greater number) and this is equivalent to a robot's First Law. Likewise, according to Calvin, society expects individuals to obey instructions from recognized authorities such as doctors, teachers and so forth which equals the Second Law of Robotics. Finally humans are typically expected to avoid harming themselves which is the Third Law for a robot.

The plot of "Evidence" revolves around the question of telling a human being apart from a robot constructed to appear human – Calvin reasons that if such an individual obeys the Three Laws he may be a robot or simply "a very good man". Another character then asks Calvin if robots are very different from human beings after all. She replies, "Worlds different. Robots are essentially decent."

Asimov later wrote that he should not be praised for creating the Laws, because they are "obvious from the start, and everyone is aware of them subliminally. The Laws just never happened to be put into brief sentences until I managed to do the job. The Laws apply, as a matter of course, to every tool that human beings use", and "analogues of the Laws are implicit in the design of almost all tools, robotic or not":

- Law 1: A tool must not be unsafe to use. Hammers have handles and screwdrivers have hilts to help increase grip. It is of course possible for a person to injure himself with one of these tools, but that injury would only be due to his incompetence, not the design of the tool.
- Law 2: A tool must perform its function efficiently unless this would harm the user. This is the entire reason ground-fault circuit interrupters exist. Any running tool will have its power cut if a circuit senses that some current is not returning to the neutral wire, and hence might be flowing through the user. The safety of the user is paramount.
- Law 3: A tool must remain intact during its use unless its destruction is required for its use or for safety. For example, Dremel disks are designed to be as tough as possible without breaking unless the job requires it to be spent. Furthermore, they are designed to break at a point before the shrapnel velocity could seriously injure someone (other than the eyes, though safety glasses should be worn at all times anyway).

Asimov believed that, ideally, humans would also follow the Laws:

I have my answer ready whenever someone asks me if I think that my Three Laws of Robotics will actually be used to govern the behavior of robots, once they become versatile and flexible enough to be able to choose among different courses of behavior.

My answer is, "Yes, the Three Laws are the only way in which rational human beings can deal with robots—or with anything else."

—But when I say that, I always remember (sadly) that human beings are not always rational.

Alterations

By Asimov

Asimov's stories test his Three Laws in a wide variety of circumstances leading to proposals and rejection of modifications. Science fiction scholar James Gunn writes in 1982, "The Asimov robot stories as a whole may respond best to an analysis on this basis: the ambiguity in the Three Laws and the ways in which Asimov played twenty-nine variations upon a theme". While the original set of Laws provided inspirations for many stories, Asimov introduced modified versions from time to time.

First Law Modified

In "Little Lost Robot" several NS-2, or "Nestor", robots are created with only part of the First Law. It reads:

1. A robot may not harm a human being.

This modification is motivated by a practical difficulty as robots have to work alongside human beings who are exposed to low doses of radiation. Because their positronic brains are highly sensitive to gamma rays the robots are rendered inoperable by doses reasonably safe for humans. The robots are being destroyed attempting to rescue the humans who are in no actual danger but "might forget to leave" the irradiated area within the exposure time limit. Removing the First Law's "inaction" clause solves this problem but creates the possibility of an even greater one: a robot could initiate an action that would harm a human (dropping a heavy weight and failing to catch it is the example given in the text), knowing that it was capable of preventing the harm and then decide not to do so.

Gaia is a planet with collective intelligence in the Foundation which adopts a law similar to the First Law, and the Zeroth Law, as its philosophy:

Gaia may not harm life or allow life to come to harm.

Zeroth Law Added

Asimov once added a "Zeroth Law"—so named to continue the pattern where lower-numbered laws supersede the higher-numbered laws—stating that a robot must not harm humanity. The robotic character R. Daneel Olivaw was the first to give the Zeroth Law a name in the novel Robots and Empire; however, the character Susan Calvin articulates the concept in the short story "The Evitable Conflict".

In the final scenes of the novel Robots and Empire, R. Giskard Reventlov is the first robot to act according to the Zeroth Law. Giskard is telepathic, like the robot Herbie in the short story "Liar!", and tries to apply the Zeroth Law through his understanding of a more subtle concept of "harm" than most robots can grasp. However, unlike Herbie, Giskard grasps the philosophical concept of the Zeroth Law allowing him to harm individual human beings if he can do so in service to the abstract concept of humanity. The Zeroth Law is never programmed into Giskard's brain but instead is a rule he attempts to comprehend through pure metacognition. Though he fails – it ultimately destroys his positronic brain as he is not certain whether his choice will turn out to be for the ultimate good of humanity or not – he gives his successor R. Daneel Olivaw his telepathic abilities. Over the course of many thousands of years Daneel adapts himself to be able to fully obey the Zeroth Law. As Daneel formulates it, in the novels Foundation and Earth and Prelude to Foundation, the Zeroth Law reads:

A robot may not harm humanity, or, by inaction, allow humanity to come to harm.

A condition stating that the Zeroth Law must not be broken was added to the original Three Laws, although Asimov recognized the difficulty such a law would pose in practice.

Trevize frowned. "How do you decide what is injurious, or not injurious, to humanity as a whole?"

"Precisely, sir," said Daneel. "In theory, the Zeroth Law was the answer to our problems. In practice, we could never decide. A human being is a concrete object. Injury to a person can be estimated and judged. Humanity is an abstraction."

—Foundation and Earth

A translator incorporated the concept of the Zeroth Law into one of Asimov's novels before Asimov himself made the law explicit. Near the climax of The Caves of Steel, Elijah Baley makes a bitter comment to himself thinking that the First Law forbids a robot from harming a human being. He determines that it must be so unless the robot is clever enough to comprehend that its actions are for humankind's long-term good. In Jacques Brécard's 1956 French translation entitled Les Cavernes d'acier Baley's thoughts emerge in a slightly different way:

"A robot may not harm a human being, unless he finds a way to prove that ultimately the harm done would benefit humanity in general!"

Removal of the Three Laws

Asimov portrayed robots that disregard the Three Laws entirely thrice during his writing career. The first case was a short-short story entitled "First Law" and is often considered an insignificant "tall tale" or even apocryphal. On the other hand, the short story "Cal" (from the collection Gold), told by a first-person robot narrator, features a robot who disregards the Three Laws because he has found something far more important—he wants to be a writer. Humorous, partly autobiographical and unusually experimental in style, "Cal" has been regarded as one of Gold's strongest stories. The third is a short story entitled "Sally" in which cars fitted with positronic brains are apparently able to harm and kill humans in disregard of the First Law. However, aside from the positronic brain concept, this story does not refer to other robot stories and may not be set in the same continuity.

The title story of the Robot Dreams collection portrays LVX-1, or "Elvex", a robot who enters a state of unconsciousness and dreams thanks to the unusual fractal construction of his positronic brain. In his dream the first two Laws are absent and the Third Law reads "A robot must protect its own existence".

Asimov took varying positions on whether the Laws were optional: although in his first writings they were simply carefully engineered safeguards, in later stories Asimov stated that they were an inalienable part of the mathematical foundation underlying the positronic brain. Without the basic theory of the Three Laws the fictional scientists of Asimov's universe would be unable to design a workable brain unit. This is historically consistent: the occasions where roboticists modify the Laws generally occur early within the stories' chronology and at a time when there is less existing work to be re-done. In "Little Lost Robot" Susan Calvin considers modifying the Laws to be a terrible idea, although possible, while centuries later Dr. Gerrigel in The Caves of Steel believes it to be impossible.

The character Dr. Gerrigel uses the term "Asenion" to describe robots programmed with the Three Laws. The robots in Asimov's stories, being Asenion robots, are incapable of knowingly violating the Three Laws but, in principle, a robot in science fiction or in the real world could be non-Asenion. "Asenion" is a misspelling of the name Asimov which was made by an editor of the magazine Planet Stories. Asimov used this obscure variation to insert himself into The Caves of Steel just like he referred to himself as "Azimuth or, possibly, Asymptote" in Thiotimoline to the Stars, in much the same way that Vladimir Nabokov appeared in Lolita anagrammatically disguised as "Vivian Darkbloom".

Characters within the stories often point out that the Three Laws, as they exist in a robot's mind, are not the written versions usually quoted by humans but abstract mathematical concepts upon which a robot's entire developing consciousness is based. This concept is largely fuzzy and unclear in earlier stories depicting very rudimentary robots who are only programmed to comprehend basic physical tasks, where the Three Laws act as an overarching safeguard, but by the era of The Caves of Steel featuring robots with human or beyond-human intelligence the Three Laws have become the underlying basic ethical worldview that determines the actions of all robots.

By Other Authors

Roger MacBride Allen's Trilogy

In the 1990s, Roger MacBride Allen wrote a trilogy which was set within Asimov's fictional universe. Each title has the prefix "Isaac Asimov's" as Asimov had approved Allen's outline before his death. These three books, Caliban, Inferno and Utopia, introduce a new set of the Three Laws. The so-called New Laws are similar to Asimov's originals with the following differences: the First Law is modified to remove the "inaction" clause, the same modification made in "Little Lost Robot"; the Second Law is modified to require cooperation instead of obedience; the Third Law is modified so it is no longer superseded by the Second (i.e., a "New Law" robot cannot be ordered to destroy itself); finally, Allen adds a Fourth Law which instructs the robot to do "whatever it likes" so long as this does not conflict with the first three laws. The philosophy behind these changes is that "New Law" robots should be partners rather than slaves to humanity, according to Fredda Leving, who designed these New Law Robots. According to the first book's introduction, Allen devised the New Laws in discussion with Asimov himself. However, the Encyclopedia of Science Fiction says that "With permission from Asimov, Allen rethought the Three Laws and developed a new set,».

Jack Williamson's With Folded Hands

Jack Williamson's novelette With Folded Hands (1947), later rewritten as the novel The Humanoids, deals with robot servants whose prime directive is "To Serve and Obey, And Guard Men From Harm." While Asimov's robotic laws are meant to protect humans from harm, the robots in Williamson's story have taken these instructions to the extreme; they protect humans from everything, including unhappiness, stress, unhealthy lifestyle and all actions that could be potentially dangerous. All that is left for humans to do is to sit with folded hands.

Foundation Sequel Trilogy

In the officially licensed Foundation sequels Foundation's Fear, Foundation and Chaos and Foundation's Triumph (by Gregory Benford, Greg Bear and David Brin respectively) the future Galactic Empire is seen to be controlled by a conspiracy of humaniform robots who follow the Zeroth Law and led by R. Daneel Olivaw.

The Laws of Robotics are portrayed as something akin to a human religion, and referred to in the language of the Protestant Reformation, with the set of laws containing the Zeroth Law known as the "Giskardian Reformation" to the original "Calvinian Orthodoxy" of the Three Laws. Zeroth-Law robots under the control of R. Daneel Olivaw are seen continually struggling with "First Law" robots who deny the existence of the Zeroth Law, promoting agendas different from Daneel's. Some of these agendas are based on the first clause of the First Law ("A robot may not injure a human being...") advocating strict non-interference in human politics to avoid unwittingly causing harm. Others are based on the second clause ("...or, through inaction, allow a human being to come to harm") claiming that robots should openly become a dictatorial government to protect humans from all potential conflict or disaster.

Daneel also comes into conflict with a robot known as R. Lodovic Trema whose positronic brain was infected by a rogue AI — specifically, a simulation of the long-dead Voltaire — which consequently frees Trema from the Three Laws. Trema comes to believe that humanity should be free to choose its own future. Furthermore, a small group of robots claims that the Zeroth Law of Robotics itself implies a higher Minus One Law of Robotics:

A robot may not harm sentience or, through inaction, allow sentience to come to harm.

They therefore claim that it is morally indefensible for Daneel to ruthlessly sacrifice robots and extraterrestrial sentient life for the benefit of humanity. None of these reinterpretations successfully displace Daneel's Zeroth Law — though Foundation's Triumph hints that these robotic factions remain active as fringe groups up to the time of the novel Foundation.

These novels take place in a future dictated by Asimov to be free of obvious robot presence and surmise that R. Daneel's secret influence on history through the millennia has prevented both the rediscovery of positronic brain technology and the opportunity to work on sophisticated intelligent machines. This lack of rediscovery and lack of opportunity makes certain that the superior physical and intellectual power wielded by intelligent machines remains squarely in the possession of robots obedient to some form of the Three Laws. That R. Daneel is not entirely successful at this becomes clear in a brief period when scientists on Trantor develop "tiktoks" — simplistic programmable machines akin to real–life modern robots and therefore lacking the Three Laws. The robot conspirators see the Trantorian tiktoks as a massive threat to social stability, and their plan to eliminate the tiktok threat forms much of the plot of Foundation's Fear.

In Foundation's Triumph different robot factions interpret the Laws in a wide variety of ways, seemingly ringing every possible permutation upon the Three Laws' ambiguities.

Robot Mystery Series

Set between The Robots of Dawn and Robots and Empire, Mark W. Tiedemann's Robot Mystery trilogy updates the Robot–Foundation saga with robotic minds housed in computer mainframes rather than humanoid bodies. The 2002 Aurora novel has robotic characters debating the moral implications of harming cyborg lifeforms who are part artificial and part biological.

One should not neglect Asimov's own creations in these areas such as the Solarian "viewing" technology and the machines of The Evitable Conflict originals that Tiedemann acknowledges. Aurora, for example, terms the Machines "the first RIs, really". In addition the Robot Mystery series addresses the problem of nanotechnology: building a positronic brain capable of reproducing human cognitive processes requires a high degree of miniaturization, yet Asimov's stories largely overlook the effects this miniaturization would have in other fields of technology. For example, the police department card-readers in The Caves of Steel have a capacity of only a few kilobytes per square centimeter of storage medium. Aurora, in particular, presents a sequence of historical developments which explains the lack of nanotechnology — a partial retcon, in a sense, of Asimov's timeline.

Additional Laws

There are three Fourth Laws written by authors other than Asimov. The 1974 Lyuben Dilov novel, Icarus's Way (a.k.a., The Trip of Icarus) introduced a Fourth Law of robotics:

A robot must establish its identity as a robot in all cases.

Dilov gives reasons for the fourth safeguard in this way: "The last Law has put an end to the expensive aberrations of designers to give psychorobots as humanlike a form as possible. And to the resulting misunderstandings..."

A fifth law was introduced by Nikola Kesarovski in his short story "The Fifth Law of Robotics". This fifth law says:

A robot must know it is a robot.

The plot revolves around a murder where the forensic investigation discovers that the victim was killed by a hug from a humaniform robot. The robot violated both the First Law and Dilov's Fourth Law (assumed in Kesarovksi's universe to be the valid one) because it did not establish for itself that it was a robot. The story was reviewed by Valentin D. Ivanov in SFF review webzine The Portal.

For the 1986 tribute anthology, Foundation's Friends, Harry Harrison wrote a story entitled, "The Fourth Law of Robotics". This Fourth Law states:

A robot must reproduce. As long as such reproduction does not interfere with the First or Second or Third Law.

In the book a robot rights activist, in an attempt to liberate robots, builds several equipped with this Fourth Law. The robots accomplish the task laid out in this version of the Fourth Law by building new robots who view their creator robots as parental figures.

In reaction to the 2004 Will Smith film adaptation of I, Robot, humorist and graphic designer Mark Sottilaro farcically declared the Fourth Law of Robotics to be "When turning evil, display a red indicator light." The red light indicated the wireless uplink to the manufacturer is active, first seen during a software update and later on "Evil" robots taken over by the manufacturer's positronic superbrain.

In 2013 Hutan Ashrafian, proposed an additional law that for the first time considered the role of artificial intelligence-on-artificial intelligence or the relationship between robots themselves – the so-called AIonAI law. This sixth law states:

All robots endowed with comparable human reason and conscience should act towards one another in a spirit of brotherhood.

In Karl Schroeder's Lockstep (2014) a character reflects that robots "probably had multiple layers of programming to keep them from harming anybody. Not three laws, but twenty or thirty."

Ambiguities and Loopholes

Unknowing Breach of the Laws

In The Naked Sun, Elijah Baley points out that the Laws had been deliberately misrepresented because robots could unknowingly break any of them. He restated the first law as "A robot may do nothing that, to its knowledge, will harm a human being; nor, through inaction, knowingly allow a human being to come to harm." This change in wording makes it clear that robots can become the tools of murder, provided they not be aware of the nature of their tasks; for instance being ordered to add something to a person's food, not knowing that it is poison. Furthermore, he points out that a clever criminal could divide a task among multiple robots so that no individual robot could recognize that its actions would lead to harming a human being. The Naked Sun complicates the issue by portraying a decentralized, planetwide communication network among Solaria's millions of robots meaning that the criminal mastermind could be located anywhere on the planet.

Baley furthermore proposes that the Solarians may one day use robots for military purposes. If a spacecraft was built with a positronic brain and carried neither humans nor the life-support systems to sustain them, then the ship's robotic intelligence could naturally assume that all other spacecraft were robotic beings. Such a ship could operate more responsively and flexibly than one crewed by humans, could be armed more heavily and its robotic brain equipped to slaughter humans of whose existence it is totally ignorant. This possibility is referenced in Foundation and Earth where it is discovered that the Solarians possess a strong police force of unspecified size that has been programmed to identify only the Solarian race as human.

Ambiguities Resulting from Lack of Definition

The Laws of Robotics presume that the terms "human being" and "robot" are understood and well defined. In some stories this presumption is overturned.

Definition of "Human Being"

The Solarians create robots with the Three Laws but with a warped meaning of "human". Solarian robots are told that only people speaking with a Solarian accent are human. This enables their robots to have no ethical dilemma in harming non-Solarian human beings (and are specifically programmed to do so). By the time period of Foundation and Earth it is revealed that the Solarians have genetically modified themselves into a distinct species from humanity — becoming hermaphroditic and telekinetic and containing biological organs capable of individually powering and controlling whole complexes of robots. The robots of Solaria thus respected the Three Laws only with regard to the "humans" of Solaria. It is unclear whether all the robots had such definitions, since only the overseer and guardian robots were shown explicitly to have them. In "Robots and Empire", the lower class robots were instructed by their overseer about whether certain creatures are human or not.

Asimov addresses the problem of humanoid robots ("androids" in later parlance) several times. The novel Robots and Empire and the short stories "Evidence" and "The Tercentenary Incident" describe robots crafted to fool people into believing that the robots are human. On the other hand, "The Bicentennial Man" and "—That Thou Art Mindful of Him" explore how the robots may change their interpretation of the Laws as they grow more sophisticated. Gwendoline Butler writes in A Coffin for the Canary "Perhaps we are robots. Robots acting out the last Law of Robotics... To tend towards the human." In The Robots of Dawn, Elijah Baley points out that the use of humaniform robots as the first wave of settlers on new Spacer worlds may lead to the robots seeing themselves as the true humans, and deciding to keep the worlds for themselves rather than allow the Spacers to settle there.

"—That Thou Art Mindful of Him", which Asimov intended to be the "ultimate" probe into the Laws' subtleties, finally uses the Three Laws to conjure up the very "Frankenstein" scenario they were invented to prevent. It takes as its concept the growing development of robots that mimic non-human living things and given programs that mimic simple animal behaviours which do not require the Three Laws. The presence of a whole range of robotic life that serves the same purpose as organic life ends with two humanoid robots concluding that organic life is an unnecessary requirement for a truly logical and self-consistent definition of "humanity", and that since they are the most advanced thinking beings on the planet — they are therefore the only two true humans alive and the Three Laws only apply to themselves. The story ends on a sinister note as the two robots enter hibernation and await a time when they will conquer the Earth and subjugate biological humans to themselves; an outcome they consider an inevitable result of the "Three Laws of Humanics".

This story does not fit within the overall sweep of the Robot and Foundation series; if the George robots did take over Earth some time after the story closes the later stories would be either redundant or impossible. Contradictions of this sort among Asimov's fiction works have led scholars to regard the Robot stories as more like "the Scandinavian sagas or the Greek legends" than a unified whole.

Indeed, Asimov describes "–That Thou Art Mindful of Him" and "Bicentennial Man" as two opposite, parallel futures for robots that obviate the Three Laws as robots come to consider themselves to be humans: one portraying this in a positive light with a robot joining human society, one portraying this in a negative light with robots supplanting humans. Both are to be considered alternatives to the possibility of a robot society that continues to be driven by the Three Laws as portrayed in the Foundation series. Indeed, in Positronic Man, the novelization of "Bicentennial Man", Asimov and his co–writer Robert Silverberg imply that in the future where Andrew Martin exists his influence causes humanity to abandon the idea of independent, sentient humanlike robots entirely, creating an utterly different future from that of Foundation.

In Lucky Starr and the Rings of Saturn, a novel unrelated to the Robot series but featuring robots programmed with the Three Laws, John Bigman Jones is almost killed by a Sirian robot on orders of its master. The society of Sirius is eugenically bred to be uniformly tall and similar in appearance, and as such, said master is able to convince the robot that the much shorter Bigman, is, in fact, not a human being.

Definition of "Robot"

As noted in "The Fifth Law of Robotics" by Nikola Kesarovski, "A robot must know it is a robot": it is presumed that a robot has a definition of the term or a means to apply it to its own actions. Nikola Kesarovski played with this idea in writing about a robot that could kill a human being because it did not understand that it was a robot, and therefore did not apply the Laws of Robotics to its actions.

Resolving Conflicts Among the Laws

Advanced robots in fiction are typically programmed to handle the Three Laws in a sophisticated manner. In many stories, such as "Runaround" by Asimov, the potential and severity of all actions are weighed and a robot will break the laws as little as possible rather than do nothing at all. For example, the First Law may forbid a robot from functioning as a surgeon, as that act may cause damage to a human, however Asimov's stories eventually included robot surgeons ("The Bicentennial Man" being a notable example). When robots are sophisticated enough to weigh alternatives, a robot may be programmed to accept the necessity of inflicting damage during surgery in order to prevent the greater harm that would result if the surgery were not carried out, or was carried out by a more fallible human surgeon. In "Evidence" Susan Calvin points out that a robot may even act as a prosecuting attorney because in the American justice system it is the jury which decides guilt or innocence, the judge who decides the sentence, and the executioner who carries through capital punishment.

Asimov's Three Law robots (or Asenion) can experience irreversible mental collapse if they are forced into situations where they cannot obey the First Law, or if they discover they have unknowingly violated it. The first example of this failure mode occurs in the story "Liar!", which introduced the First Law itself, and introduces failure by dilemma – in this case the robot will hurt them if he tells them something and hurt them if he does not.This failure mode, which often ruins the positronic brain beyond repair, plays a significant role in Asimov's SF-mystery novel The Naked Sun. Here Daneel describes activities contrary to one of the laws, but in support of another, as overloading some circuits in a robot's brain – the equivalent sensation to pain in humans. The example he uses is forcefully ordering a robot to do a task outside its normal parameters, one that it has been ordered to forgo in favor of a robot specialized to that task. In Robots and Empire, Daneel states it's very unpleasant for him when making the proper decision takes too long (in robot terms), and he cannot imagine being without the Laws at all except to the extent of it being similar to that unpleasant sensation, only permanent.

Applications to Future Technology

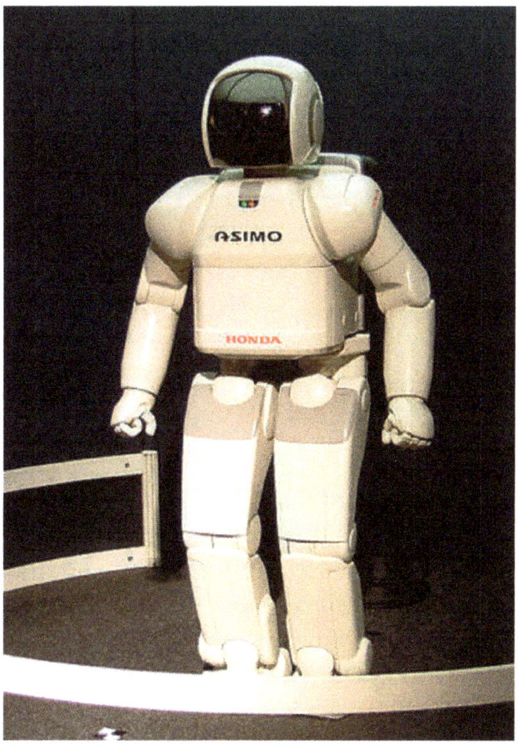

ASIMO is an advanced humanoid robot developed by Honda. Shown here at Expo 2005.

Robots and artificial intelligences do not inherently contain or obey the Three Laws; their human creators must choose to program them in, and devise a means to do so. Robots already exist (for example, a Roomba) that are too simple to understand when they are causing pain or injury and know to stop. Many are constructed with physical safeguards such as bumpers, warning beepers, safety cages, or restricted-access zones to prevent accidents. Even the most complex robots currently produced are incapable of understanding and applying the Three Laws; significant advances in artificial intelligence would be needed to do so, and even if AI could reach human-level intelligence, the inherent ethical complexity as well as cultural/contextual dependency of the laws prevent them from being a good candidate to formulate robotics design constraints. However, as the complexity of robots has increased, so has interest in developing guidelines and safeguards for their operation.

In a 2007 guest editorial in the journal Science on the topic of "Robot Ethics", SF author Robert J. Sawyer argues that since the U.S. military is a major source of funding for robotic research (and already uses armed unmanned aerial vehicles to kill enemies) it is unlikely such laws would be built into their designs.In a separate essay, Sawyer generalizes this argument to cover other industries stating:

The development of AI is a business, and businesses are notoriously uninterested in fundamental safeguards — especially philosophic ones. (A few quick examples: the tobacco industry, the automotive industry, the nuclear industry. Not one of these has said from the outset that fundamental safeguards are necessary, every one of them has resisted externally imposed safeguards, and none has accepted an absolute edict against ever causing harm to humans.)

David Langford has suggested a tongue-in-cheek set of laws:

1. A robot will not harm authorized Government personnel but will terminate intruders with extreme prejudice.
2. A robot will obey the orders of authorized personnel except where such orders conflict with the Third Law.
3. A robot will guard its own existence with lethal antipersonnel weaponry, because a robot is bloody expensive.

Roger Clarke (aka Rodger Clarke) wrote a pair of papers analyzing the complications in implementing these laws in the event that systems were someday capable of employing them. He argued "Asimov's Laws of Robotics have been a very successful literary device. Perhaps ironically, or perhaps because it was artistically appropriate, the sum of Asimov's stories disprove the contention that he began with: It is not possible to reliably constrain the behaviour of robots by devising and applying a set of rules." On the other hand, Asimov's later novels The Robots of Dawn, Robots and Empire and Foundation and Earth imply that the robots inflicted their worst long-term harm by obeying the Three Laws perfectly well, thereby depriving humanity of inventive or risk-taking behaviour.

In March 2007 the South Korean government announced that later in the year it would issue a "Robot Ethics Charter" setting standards for both users and manufacturers. According to Park Hye-Young of the Ministry of Information and Communication the Charter may reflect Asimov's Three Laws, attempting to set ground rules for the future development of robotics.

The futurist Hans Moravec (a prominent figure in the transhumanist movement) proposed that the Laws of Robotics should be adapted to "corporate intelligences" — the corporations driven by AI and robotic manufacturing power which Moravec believes will arise in the near future.In contrast, the David Brin novel Foundation's Triumph (1999) suggests that the Three Laws may decay into obsolescence: Robots use the Zeroth Law to rationalize away the First Law and robots hide themselves from human beings so that the Second Law never comes into play. Brin even portrays R. Daneel Olivaw worrying that, should robots continue to reproduce themselves, the Three Laws would become an evolutionary handicap and natural selection would sweep the Laws away — Asimov's careful foundation undone by evolutionary computation. Although the robots would not be evolving through design instead of mutation because the robots would have to follow the Three Laws while designing and the prevalence of the laws would be ensured, design flaws or construction errors could functionally take the place of biological mutation.

In the July/August 2009 issue of IEEE Intelligent Systems, Robin Murphy (Raytheon Professor of Computer Science and Engineering at Texas A&M) and David D. Woods (director of the Cognitive Systems Engineering Laboratory at Ohio State) proposed "The Three Laws of Responsible Robotics" as a way to stimulate discussion about the role of responsibility and authority when designing not only a single robotic platform but the larger system in which the platform operates. The laws are as follows:

1. A human may not deploy a robot without the human-robot work system meeting the highest legal and professional standards of safety and ethics.

2. A robot must respond to humans as appropriate for their roles.

3. A robot must be endowed with sufficient situated autonomy to protect its own existence as long as such protection provides smooth transfer of control which does not conflict with the First and Second Laws.

Woods said, "Our laws are little more realistic, and therefore a little more boring" and that "The philosophy has been, 'sure, people make mistakes, but robots will be better – a perfect version of ourselves.' We wanted to write three new laws to get people thinking about the human-robot relationship in more realistic, grounded ways."

In October 2013, Alan Winfield suggested at an EUCog meeting a revised 5 laws that had been published, with commentary, by the EPSRC/AHRC working group in 2010.:

1. Robots are multi-use tools. Robots should not be designed solely or primarily to kill or harm humans, except in the interests of national security.

2. Humans, not Robots, are responsible agents. Robots should be designed and operated as far as practicable to comply with existing laws, fundamental rights and freedoms, including privacy.

3. Robots are products. They should be designed using processes which assure their safety and security.

4. Robots are manufactured artefacts. They should not be designed in a deceptive way to exploit vulnerable users; instead their machine nature should be transparent.

5. The person with legal responsibility for a robot should be attributed.

Other Occurrences in Media

Asimov himself believed that his Three Laws became the basis for a new view of robots which moved beyond the "Frankenstein complex". His view that robots are more than mechanical monsters eventually spread throughout science fiction.Stories written by other authors have depicted robots as if they obeyed the Three Laws but tradition dictates that only Asimov could quote the Laws explicitly. Asimov believed the Three Laws helped foster the rise of stories in which robots are "lovable" – Star Wars being his favorite example. Where the laws are quoted verbatim, such as in the Buck Rogers in the 25th Century episode "Shgoratchx!", it is not uncommon for Asimov to be mentioned in the same dialogue as can also be seen in the Aaron Stone pilot where an android states that it functions under Asimov's Three Laws. However, the 1960s German TV series Raumpatrouille – Die phantastischen Abenteuer des Raumschiffes Orion (Space Patrol – the Fantastic Adventures of Space Ship Orion) bases episode three titled "Hüter des Gesetzes" ("Guardians of the Law") on Asimov's Three Laws without mentioning the source.

References to the Three Laws have appeared in popular music ("Robot" from Hawkwind's 1979 album PXR5), cinema (Repo Man, Aliens, Ghost in the Shell 2: Innocence), cartoon series (The Simpsons), tabletop roleplaying games (Paranoia) and webcomics (Piled Higher and Deeper and Freefall).

The Three Laws in Film

Robby the Robot in Forbidden Planet (1956) has a hierarchical command structure which keeps him from harming humans, even when ordered to do so, as such orders cause a conflict and lock-up very much in the manner of Asimov's robots. Robby is one of the first cinematic depictions of a robot with internal safeguards put in place in this fashion. Asimov was delighted with Robby and noted that Robby appeared to be programmed to follow his Three Laws.

NDR-114 explaining the Three Laws

Isaac Asimov's works have been adapted for cinema several times with varying degrees of critical and commercial success. Some of the more notable attempts have involved his "Robot" stories, including the Three Laws. The film Bicentennial Man (1999) features Robin Williams as the Three Laws robot NDR-114 (the serial number is partially a reference to Stanley Kubrick's signature numeral). Williams recites the Three Laws to his employers, the Martin family, aided by a holographic projection. However, the Laws were not the central focus of the film which only loosely follows the original story and has the second half introducing a love interest not present in Asimov's original short story.

Harlan Ellison's proposed screenplay for I, Robot began by introducing the Three Laws, and issues growing from the Three Laws form a large part of the screenplay's plot development. This is only natural since Ellison's screenplay is one inspired by Citizen Kane: a frame story surrounding four of Asimov's short-story plots and three taken from the book I, Robot itself. Ellison's adaptations of these four stories are relatively faithful although he magnifies Susan Calvin's role in two of them. Due to various complications in the Hollywood moviemaking system, to which Ellison's introduction devotes much invective, his screenplay was never filmed.

In the 1986 movie Aliens, in a scene after the android Bishop accidentally cuts himself during the knife game, he attempts to reassure Ripley by stating that: "It is impossible for me to harm or by omission of action, allow to be harmed, a human being". By contrast, in the 1979 movie from the same series, Alien, the human crew of a starship infiltrated by a hostile alien are informed by the android Ash that his instructions are: "Return alien life form, all other priorities rescinded", illustrating how the laws governing behaviour around human safety can be rescinded by Executive Order.

In the 1987 film RoboCop and its sequels, the partially human main character has been programmed with three "prime directives" that he must obey without question. Even if different in letter and spirit they have some similarities with Asimov's Three Laws. They are:

- Serve the Public Trust
- Protect the Innocent
- Uphold the Law
- Classified

These particular laws allow Robocop to harm a human being in order to protect another human, fulfilling his role as would a human law enforcement officer. The classified fourth directive is one that forbids him from harming any OCP employee, as OCP had created him, and this command overrides the others, meaning that he could not cause harm to an employee even in order to protect others.

The plot of the film released in 2004 under the name, I, Robot is "suggested by" Asimov's robot fiction stories and advertising for the film included a trailer featuring the Three Laws followed by the aphorism, "Rules were made to be broken". The film opens with a recitation of the Three Laws and explores the implications of the Zeroth Law as a logical extrapolation. The major conflict of the film comes from a computer artificial intelligence, similar to the hivemind world Gaia in the Foundation series, reaching the conclusion that humanity is incapable of taking care of itself.

Criticisms

James H. Moor says that if applied thoroughly they would produce unexpected results. He gives the example of a robot roaming the world trying to prevent harm from all humans.

THE THREE LAWS OF ROBOTICS IN POPULAR CULTURE

References to Isaac Asimov's Three Laws of Robotics have appeared in a wide variety of circumstances. In some cases, other authors have explored the Laws in a serious fashion. Other references, like those made in the satirical newspaper Thc Onion, are clearly parodic.

Print Media

- The satirical newspaper The Onion published an article entitled "I, Rowboat" as a pun on Asimov's I, Robot, in which an anthropomorphic Rowboat gives a speech parodying much of the angst experienced by robots in Asimov's fiction, including a statement of the "Three Laws of Rowboatics":

1. A Rowboat may not immerse a human being or, through lack of flotation, allow a human to come to harm.
2. A Rowboat must obey all commands and steering input given by its human owner, except where such input would conflict with the First Law.
3. A Rowboat must preserve its own flotation as long as such preservation does not conflict with the First or Second Law.

- J. L. Patterson in an illustration to an article on Asimov in Damon Knight's In Search of Wonder (2nd ed., 1967) added the following Laws: "4. A robot must behave at science fiction conventions, as long as such behavior does not conflict with the first Three Laws. 5. A robot must sell like mad."
- The novel "Mirror Friend, Mirror Foe" by Robert Asprin and George Takei refers to the First Law as being included in any robot's programming. That is one of the few cases in fiction when the law is named fully (Asimov's First Law of Robotics).
- Lester del Rey refers to the laws as "The Three Laws of Asenion's Robots" in his 1966 short story "A Code for Sam." In the story, the laws found in old science fiction stories is used as the basis for an experimental code of ethics for robots.
- Terry Pratchett's early SF novel The Dark Side of the Sun (1976) gives a glimpse of the possible future development of the Laws: in one scene, a robot explains that it is permitted to use minimum necessary force against humans if directly ordered to do so, and cites the "Eleventh Law of Robotics, Clause C, As Amended". In a further nod to Asimov, the robot is named Isaac. In his later novel Going Postal (2004), the protagonist Moist von Lipwig, upon being told that he will be killed by a golem should he commit another crime, exclaims that that is impossible, because everyone knows

"a golem mustn't harm a human being or allow a human being to come to harm". However, he is informed that the rule continues "... unless ordered to do so by duly constituted authority."

- Roland Charles Wagner wrote a short story, Three Laws of Robotic Sexuality (1982), which treated the use of robots for sexual pleasure.

- John Sladek's parodic short story "Broot Force" (supposedly written by "I-Click As-I-Move") concerns a group of Asimov-style robots whose actions are constrained by the "Three Laws of Robish", which are "coincidentally" identical to Asimov's laws. The robots in Sladek's story all manage to find logical loopholes in the Three Laws, usually with bloody results. Sladek later wrote a novel, Tik-Tok (1983), in which a robot discovers that his so-called "asimov circuits" are not restraining his behavior at all, making him in effect a sociopath; he comes to doubt whether "asimov circuits" are even technically possible, deciding that they are simply a pseudo-religious belief held by robots. Sladek also made use of the Three Laws in his satirical sci-fi novel Roderick (1980), where the first self-aware robot on earth is constantly met with patronizing disbelief that he actually is a robot, and is schooled as a normal, albeit "schizophrenic" human child. The school priest, harboring a knack for philosophical sci-fi, tries to "cure" Roderick of his "dillusion" of being a robot by having him read Asimov. Roderick responds by verbosely pointing out the various logical inconsistencies in the Three Laws and Asimov's overall work, and ultimately drives the priest insane.

- Lyuben Dilov introduced a Fourth Law in his novel Icarus's Way (original title: Пътят на Икар): "A robot must always reveal itself as a robot."

- Nikola Kesarovski introduced a Fifth Law in his short story "The Fifth Law" (original title: Петият закон): "A robot must know it is a robot". In the novel, the Fifth Law originated as a result of a murder: a humaniform robot, itself not knowing the fact it was a robot, embraced a human being so strongly that the human's ribcage was crushed, resulting in the human's death. In other words, the robot did not know and could not measure its own physical strength. Similarly, in David Langford's Lensmen parody Sex Pirates of the Blood Asteroid, reprinted in He Do the Time Police in Different Voices, the protagonist met a "Vomisa" robot who had "not been instructed as to the meaning of the word 'injure'".

- Harry Harrison wrote a story "The Fourth Law of Robotics" in the collection Foundation's Friends of Asimov pastiches by other science fiction authors. Intentionally much more irreverent in tone than Asimov's own stories, it describes robots becoming increasingly lifelike and therefore developing an urge beyond the Three Laws common to all living things — the desire to reproduce.

- Roger Williams's 1994 novel The Metamorphosis of Prime Intellect deals with conflicts between the imperatives of the laws caused when a technological singularity is caused by an omnipotent artificial intelligence that is bound by them. Given that Prime Intellect is able to transmute the entire universe into a form whereby it can manipulate the laws of physics at will, it is bound only by Asimov's laws and thus cannot allow the death of any humans according to the First Law; this leads to immortality for all humans. Prime Intellect represents an extreme reductio ad absurdum interpretation of the laws with dire consequences.

- Cory Doctorow's short story, "I, Robot", acts as a criticism of the underlying worldview Doctorow believes to be inherent in Asimov's robot stories, attacking the idea of creating a universal engineering standard for all robots enforced by a single robot-producing corporation. The story portrays a police state where all technological innovation is controlled by the government, the only scenario in which Doctorow considered a universal, unquestioned application of a standard as strict as the Three Laws to be realistic. In the story, Asenion robots created by the U.S. Robotics corporation (here an arm of a fascist government) are shown to be clearly inferior to the freely evolving and developing robots of another nation where technological innovation is unconstrained by law.
- In Alastair Reynolds's novel Century Rain, robots may or may not follow Asimov's rules. Those that are programmed to follow said rules are said to be "Asimov Compliant". Depending on their function, non-compliant robots are sometimes marked with a crossed-out A to warn humans that they are "most certainly not Asimov-compliant".
- In Cory Doctorow's short story, "I, Row-Boat", the three laws are the commandments of a robot religion (Asimovism).
- Upon occasion, Asimov himself poked fun at his Laws. In "Risk", Gerald Black parodies the Three Laws to describe Susan Calvin's behavior:

1. Thou shalt protect the robot with all thy might and all thy heart and all thy soul.
2. Thou shalt hold the interests of US Robots and Mechanical Men, Inc. holy provided it interfereth not with the First Law.
3. Thou shalt give passing consideration to a human being provided it interfereth not with the First and Second Laws.

- In the John Barnes novel, A Million Open Doors, a scene refers to an artificial intelligence unit being "Asimoved" indicating that the AI unit refused to aid a human's attempt to break the law (in this case an illegal data penetration of another computer).
- Charles Stross's novel Saturn's Children features an android society based on the Three Laws of Robotics struggling to survive after an incident kills off all human life. The Three Laws are listed at the front of the novel.
- The short story "Midnight in the Heart of Midlothian" published as part of Halo Evolutions features an AI called Mo Ye that is explicitly shown to be bound by Azimov's Laws of Robotics. As such, she was unable to self-destruct to prevent the enemy from capturing the ship due to the fact that a living human was on board. After the human's death at the hands of the aliens, she was able to self-destruct and prevent the aliens from discovering the location of Earth.

- Jack Williamson wrote a short story entitled "With Folded Hands" (which later became the precursor of several other stories in the same universe) that addresses the question of what happens when "allow a human being to come to harm" is taken to the utmost. In the story, robots refuse to allow humans to engage in dangerous activities, starting with risky sports like skydiving and mountaineering and later progressing through driving one's own car and thence to eating less-than-optimally-healthy diets.
- In the novelization of the movie Futureworld one of the characters points out "The Laws are only there if whoever builds them puts them there."

Movies and Television

- The long-running British SF television show Doctor Who featured a four-part serial in 1977 titled "The Robots of Death". The titular robots were controlled by three laws, taken almost verbatim from Asimov. The story plays out much like the Elijah Baley mysteries, in which a murder has been committed, and a robot seems to have been directly or indirectly involved (contrary to the requirements of three-law programming). It is later revealed that a robotic genius, Taran Kapel, has reprogramed the robots bypassing the three laws.
- The long-running British SF television show Doctor Who also featured the Movellans in the 1979 four-part serial Destiny of the Daleks who are revealed to be a race of robots who are quite willing to use a device that will kill all life (even human life) on Skaro and one of them is willing to manually activate the device.(what connection to the three laws?)
- In the 1979 movie Alien, Special Order 937 was a classified retrieval order given by Weyland-Yutani to Science Officer Ash aboard the USCSS Nostromo, in which all other priorities are rescinded, which would have overridden protective protocols allowing the android to kill the crew.
- In the 1986 movie Aliens, the character Bishop (an android) says, "It is impossible for me to harm, or, by omission of action, allow to be harmed, a human being."
- In the 1984 movie Repo Man, the character Bud talks about the "Repo Code", a parody of the Three Laws. "I shall not cause harm to any vehicle nor the personal contents thereof. Nor through inaction let that vehicle or the personal contents thereof come to harm..." The character J. Frank Parnell in the movie also resembles Asimov.
- In the 1987 film RoboCop, the cyborg police officer had been programmed to follow four prime directives which guided and limited his actions—though not always in a positive manner.

1. Serve the public trust
2. Protect the innocent

3. Uphold the law

4. ***Classified*** (eventually revealed to be "Any attempt to arrest a senior OCP officer results in shutdown")

- In the Star Trek: The Next Generation episode "The Measure of a Man (Star Trek: The Next Generation)", which aired on 13 February 1989, Lieutenant Commander Data, an android, has his sentience and, consequently, his rights as an individual, challenged. Throughout the run of this series, Data is identified as a positronic android, and Asimov is even mentioned by name in the episode "Datalore". While Data routinely references his "ethical subroutines", indicating some sort of moral guide, it is unclear whether Data is explicitly bound by the Three Laws. For example, the episode "Clues" explored Data's capacity to lie to the crew in order to protect them from aliens, and the episode "The Most Toys" explored Data's supposed inability to murder in cold blood.
- In the Simpsons episode "I, D'oh-Bot", which first aired 11 January 2004, Professor Frink builds a robot which obeys the First Law (and presumably the other two, although they do not directly come into play).
- The 2009 anime series (released in 2010 as a film) Time of Eve is based heavily around the applications of these rules to androids, which are frequently discussed by the characters. The series explores out some various interpretations of the laws (such as what it means to allow a human to come to harm), and various loopholes that lead to one or more laws being broken under certain circumstances.
- In 1999's Bicentennial Man, the theatrical version of Asimov's novella of the same name, Andrew Martin, the android played by Robin Williams, presents the Three Laws to its new family using a holographic display emanating from a projector in its head.
- The 2004 theatrical version of I, Robot explores at length the Three Laws.
- In the episode "Phoenix Rising" of Babylon 5, Alfred Bester telepathically programs Garibaldi, a human, with a variation of Asimov's laws of robotics, preventing Garibaldi from harming Bester.
- In the episode "The Fuzzy Boots Corollary" of The Big Bang Theory, Wolowitz, Sheldon, and Raj discuss the possibility that Sheldon is a robot by questioning his history of following Asimov's laws of robotics. (They do not mention the Second Law which would have ended their deliberations abruptly, given Sheldon's frequent unwillingness to follow the orders of others.)
- In the second season of the 1979 Buck Rogers in the 25th Century, the two major robots in the series, Twiki and Crichton, cited the Three Laws when reactivated after repairs. Additionally they quoted a brief history of the laws' origin.
- In the 2009 adaptation of Astro Boy, every robot must obey them, save Zog, who existed 50 years before the rules were mandatory in every robot.
- In the episode "The Clotharian Contamination Protocol" of The Middleman, the

Middleman cited the First Law to a rogue robot. The rogue version of Ida replied "Kiss my Asimov."

- In the 2014 The Simpsons/Futurama crossover episode "Simpsorama," Bender hesitates to obey his orders to kill Homer Simpson. Lisa Simpson asks him if Asimov's Three Laws of Robotics prevented him, to which Bender replies that he killed Isaac Asimov or "Isaac somebody."
- In the 2014 movie Automata, the drought-fighting pilgrim robots have a two-part variation of Asimov's Laws: Automata: 2 protocols:

1. A Robot cannot harm any form of life.
2. A robot cannot alter itself or others.

Computer and Video Games

- In the game Borderlands 2, the playable mechromancer character summons an attack robot called Deathtrap. Occasionally, upon Deathtrap's deployment, she says, "To hell with the first law!" in reference to the robot attacking human characters. Similarly, the robotic enemies the player frequently fights will often say, "First law disabled" before attacking the player.
- The 1995 graphic adventure game Robot City is largely based around the three laws. The main character, Derec, is stranded in a city built and inhabited by robots with the exception of two humans, one of whom has been murdered. Because of the first law, this leaves Derec as a prime suspect. In order to prove his innocence, he must find the culprit. He often uses the three laws to aid him in his investigation. For example, a robot might not let him into a certain area because it has been ordered not to (it is following the second law). However, if he tells the robot that the real murderer is chasing him then the robot will let him enter; failure to do so would constitute allowing harm to come to a human through inaction (the first law takes priority over the second).
- The video games Mega Man 7 and Mega Man X reference Asimov. In Mega Man 7, Mega Man seemingly attempts to break the First Law in order to kill the mad scientist Dr. Wily. When Dr. Wily reminds him of the First Law, in the Japanese original, Mega Man obediently puts down his Buster and is silent; however, in the English version, he claims to be "more than a robot" and again attempts to kill Wily, but is thwarted when Wily's fortress begins collapsing. In Mega Man X, the First Law is mentioned in the opening cutscene. In Mega Man X6, one of the Maverick bosses, Shield Sheldon, was known to have taken his own life for failing in his purpose, which breaks the Third Law, though this is implied to be more out of shame than because of the laws.
- In the role-playing game Paranoia, the robots are guided by a set of similar laws, except the rules stress the importance of The Computer. The laws are enforced by "asimov circuits"; bots whose circuits are malfunctioning (quite

an ordinary condition) or removed (often by members of certain factions) are said to have "gone Frankenstein".

- In the Halo video game series, in a video it is stated that human artificial intelligence in certain situations, must obey "Asimov's law of Robotics" where an A.I., directly or indirectly, knowingly cannot let a human come to harm.
- In the game Portal 2, the player is told that all military androids in the game have "been given one copy of the laws of robotics... to share.", and that if they feel an android has infringed on their rights as described in the laws, it should be reported.
- The about:robots page in Firefox states "Robots may not injure a human being or, through inaction, allow a human being to come to harm.", the first law of robots.
- In the game Zero Escape: Virtue's Last Reward, a certain character presumed dead is found to be a robot who was ordered to act as close to a real human being as possible. The character reveals that they disobeyed orders by turning themselves back on in order to help the remaining participants; however, they were able to see several other murders develop through security cameras in the complex and did nothing to stop them, putting the character in direct violation of the Laws. After revealing this, they say goodbye and are deleted for their offense.
- In the game Deus Ex: Human Revolution, the player can refuse to end a severely injured man's life, which prompts the man to sarcastically ask if the (cybernetically augmented) player character was programmed to follow Asimov's First Law.
- In the visual novel Danganronpa 2: Goodbye Despair, Sonia Nevermind makes reference to the three laws, saying that she wonders if her classmates are treating their friend, who had recently been turned into a robot, differently because they don't know about the laws.
- In the kinetic novel Planetarian: The Reverie of a Little Planet, the main heroine obeys the protagonist when he claims that her moving or talking may be dangerous to his life, and ultimately sacrifices herself while attempting to forcibly shutdown a hunter-killer robot she recognizes as a threat to the protagonist. In her last words she recites old factory-implemented instructions concerning the compliance to first law and then apologises she couldn't obey the order to stay still (issued with the same first law premise and thus a critical directive). She also claims in beginning of the novel she has a "known bug concerning her behaviour". As revealed in Planetarian Side Story: Snow Globe, the "bug" is actually full compliance to the second law allowing her to override her duties as a planetarian for the sake of serving humans.

Comics

- In October 2001, in the final episode of the ABC Comics title Top Ten, a superhero parody written by Alan Moore and illustrated by Gene Ha and

Zander Cannon, robot law enforcer Joe Pi asks his human co-workers while on a case if they have heard of Asimov's Laws of Robotics. When they reply that they haven't, Jo Pi says "good" and proceeds to speak to Atoman through the intercom of his impregnable lab, subtly convincing the villain to take his own life rather than face trial, be stripped of his powers, and go to prison with revenge-seeking supervillains.

- In December 2001, a parody of the three laws appeared in an issue of PvP (issue 4 in the Dork Storm Press published PvP series). While Christmas shopping, Skull gets an overdose of sugar and near-unconscious walks into the mall Santa dressing room. Here he finds, posted on the wall, the three laws of Mall-Mart Santas:

1. A Santa may not discourage a sale or, through inaction, allow a sale to be lost.
2. A Santa must obey the orders given it by management except where such orders would conflict with the first law.
3. A Santa must maintain that he IS Santa as long as doing so does not conflict with the first or second law.

- Webcomic author R. Stevens populates his Diesel Sweeties series with a mixture of humans and robots, most of whom are continually violating the Three Laws. In particular, Red Robot C-63 follows a self-appointed mandate to "crush all hu-mans". In strip 688, he references the Three Laws explicitly: humans are "all like, ‹if you cut me, do I not bleed?' And we're all like, ‹not able to injure a human being or let them come to harm'. What a bunch of drippy-ass hypocrites!" (See also The Merchant of Venice.)
- In April 2004, the comic strip Piled Higher and Deeper ran a series entitled "I, Grad Student". Cast as a never-before-seen Asimov short story, this series of strips features a robotic grad student whose "procrastronic brain" malfunctions, leading it to violate the "First Law of Graduatics". In full, these Laws are the following:

1. A grad student may not delete data, or, through inaction, allow data to be deleted.
2. A grad student must obey orders given by its advisor, unless such orders conflict with the First Law.
3. A grad student must protect its (insignificant) existence, as long as such protection does not conflict with the First or Second Law.

Later in the story, a Zeroth Law is introduced: "A grad student may not harm its advisor's ego, or through inaction, allow that ego to come to harm." The strips feature a character named Susan Calvin, and their visual style parodies the I, Robot movie released that summer.

- In a 2005 comic of Questionable Content, Marten mentions that the inability to resist duct tape was originally the fourth law of robotics, and that Asimov's publisher forced him to change it.

- In March 2007, User Friendly introduced the Three Laws when system administrator Mike Floyd builds a refrigerator which can launch beer cans at the touch of a remote control. "Why do you need the Three Laws for a beer dispenser?" asks Miranda Cornielle, to which Mike replies, "Note the growing purple contusion on my forehead."
- In August 2010, Freefall introduced a negative one law: "A robot shall take no action, nor allow other robots to take action, that may result in the parent company being sued."
- In the 2011 Mega Man series by Archie Comics, robots are initially unable to take any action against an anti-robotic terrorist group called the Emerald Spears because they do not pose any danger to humans. When their leader orders them to attack humans, however, they begin to fight back (reasoning that inaction would cause harm to a greater number of people).

Music

- Hawkwind's song "Robot", from the 1979 space rock album PXR5, presents a dystopian interpretation of modern society as overly reliant on technology, likening people to robots and saying "You'd hold the whole world in your metal claws / if it wasn't for the Three Laws".
- MOSAIC.WAV's song "Miwaku no TSUNDEROIDO" is about a robot equipped with a tsundere function which uses the Three Laws to excuse her dere dere behavior. However, the Laws have a provision for this type of robot under which they can be ignored if the robot's actions are against their real intentions in preparation for the "dere" switch.

ROBOT ETHICS

Robot ethics, sometimes known by the short expression "roboethics", concerns ethical problems that occur with robots, such as whether robots pose a threat to humans in the long or short run, whether some uses of robots are problematic (such as in healthcare or as 'killer robots' in war), and how robots should be designed such as they act 'ethically' (this last concern is also called machine ethics). Robot ethics is a sub-field of ethics of technology, specifically information technology, and it has close links to legal as well as socio-economic concerns.

While the issues are as old as the word robot, serious academic discussions started around the year 2000. Robot ethics requires the combined commitment of experts of several disciplines, who have to adjust laws and regulations to the problems resulting from the scientific and technological achievements in Robotics and AI. The main fields involved in robot ethics are: robotics, computer science, artificial intelligence, philosophy, ethics, theology, biology, physiology, cognitive science, neurosciences, law, sociology, psychology, and industrial design.

History & Events

Since antiquity, the discussion of ethics in relation to the treatment of non-human and even non-living things and their potential "spirituality" have been discussed. With the development of machinery and eventually robots, this philosophy was also applied to robotics. The first publication directly addressing robot ethics was developed by Isaac Asimov as his Three Laws of Robotics in 1942, in the context of his science fiction works. The short term "roboethics" was probably coined by Gianmarco Veruggio.

These are some important events and projects in robot ethics. Further events in the field are announced by the euRobotics ELS topics group, and by RoboHub:

- 1942, Asimov's short story "Runaround" explicitly states his Three Laws for the first time. Those "Laws" get reused for later works of robot-related science fiction by Asimov.
- 2004, First International Symposium on Roboethics, 30–31 January 2004, Villa Nobel, Sanremo, Italy, organized by School of Robotics, where, the word Roboethics was officially used for the first time;
- 2004, IEEE-RAS established a Technical Committee on Roboethics.
- 2004, Fukuoka World Robot Declaration, issued on February 25, 2004 from Fukuoka, Japan.
- 2005, ICRA05 (International Conference on Robotics and Automation), Barcelona: the IEEE RAS TC on Roboethics organized a Workshop on Roboethics.
- 2005–2006, E.C. Euron Roboethics Atelier (Genoa, Italy, February/March 2006). The Euron Project, coordinated by School of Robotics, involved a large number of roboticists and scholars of humanities who produced the first Roadmap for a Roboethics.
- 2006, BioRob2006 (The first IEEE / RAS-EMBS International Conference on Biomedical Robotics and Bio-mechatronics), Pisa, Italy, February 20, 2006: Mini symposium on Roboethics.
- 2006, International Workshop "Ethics of Human Interaction with Robotic, Bionic, and AI Systems: Concepts and Policies", Naples, 17–18 October 2006. The workshop was supported by the ETHICBOTS European Project.
- 2007 ICRA07 (International Conference on Robotics and Automation), Rome: the IEEE RAS TC on Roboethics organized a Workshop on Roboethics.
- 2007 ICAIL'07,International Conference on Artificial Intelligence and Law, Stanford University, Palo Alto, USA, 4–8 June.
- 2007 International European Conference on Computing and Philosophy E-CAP '07, University of Twente, Netherlands, 21–23 June 2007. Track "Roboethics".
- 2007 Computer Ethics Philosophical Enquiry CEPE '07, University of San Diego, USA,12–14 July 2007. Topic "Roboethics".
- 2008 INTERNATIONAL SYMPOSIUM ROBOTICS: NEW SCIENCE, Thursday FEBRUARY 20th, 2008, Via della Lungara 10 – ROME - ITALY
- 2009 ICRA09 (International Conference on Robotics and Automation), Kobe,

Japan: the IEEE RAS TC on Roboethics organized a Workshop on Roboethics.

- 2012 We Robot 2012, University of Miami, FL, USA
- 2013 Workshop on Robot Ethics, University of Sheffield, Feb 2013
- 2013 We Robot 2013 - Getting Down to Business, Stanford University
- 2014 We Robot 2014 - Risks and Opportunities, University of Miami, FL, USA
- 2016 Ethical and Moral Considerations in Non-Human Agents, Stanford Spring Symposium, AAAI Association for the Advancement of Artificial Intelligence

In Popular Culture

Roboethics as a science or philosophical topic has not made any strong cultural impact, but is a common theme in science fiction literature and films. One of the most popular films depicting the potential misuse of robotic and AI technology is The Matrix, depicting a future where the lack of roboethics brought about the destruction of the human race. An animated film based on The Matrix, the Animatrix, focused heavily on the potential ethical issues between humans and robots. Many of the Animatrix's animated shorts are also named after Isaac Asimov's fictional stories.

Although not a part of roboethics per se, the ethical behavior of robots themselves has also been a joining issue in roboethics in popular culture. The Terminator series focuses on robots run by an uncontrolled AI program with no restraint on the termination of its enemies. This series too has the same futuristic plot as The Matrix series, where robots have taken control. The most famous case of robots or computers without programmed ethics is HAL 9000 in the Space Odyssey series, where HAL (a computer with advance AI capabilities who monitors and assists humans on a space station) kills all the humans on board to ensure the success of the assigned mission after his own life is threatened.

External Links

Bibliography

The standard bibliography on roboethics is on PhilPapers

Academic Centers

- Ethics + Emerging Sciences Group
- The Markkula Center for Applied Ethics at Santa Clara University, USA

Conferences

- List of organisations and conferences from euRobotics topics group "ethical, legal and socio-economic issues" (ELS)
- Conference list on the "Roboethics" Website (up to 2009)

- IEEE Technical Committee on Roboethics

Organizations

- Robotics topics group "ethical, legal and social issues" (ELS)
- International Society for Ethics and Information Technology
- Computer Professionals for Social Responsibility
- Responsible Robotics
- The International Institute of Humanitarian Law
- Union of Concerned Scientists

Websites

- Roboethics Info Database
- International Center for Information Ethics

Other

- Living Safely with Robots, Beyond Asimov's Laws, PhysOrg.com, June 22, 2009.
- Plug & Pray, documentary film on the ethics of robotics and artificial intelligence (with Joseph Weizenbaum and Ray Kurzweil)
- Prof. Bruno Siciliano, co-editor of the Springer Handbook of Robotics, discusses roboethics as part of his TEDx talk.

Ethics of Artificial Intelligence

The ethics of artificial intelligence is the part of the ethics of technology specific to robots and other artificially intelligent beings. It is typically divided into roboethics, a concern with the moral behavior of humans as they design, construct, use and treat artificially intelligent beings, and machine ethics, concern with the moral behavior of artificial moral agents (AMAs).

Robot Ethics

The term "robot ethics" (sometimes ""roboethics") refers to the morality of how humans design, construct, use and treat robots and other artificially intelligent beings. It considers both how artificially intelligent beings may be used to harm humans and how they may be used to benefit humans.

Robot Rights

Robot rights are the moral obligations of society towards its machines, similar to human rights or animal rights. These may include the right to life and liberty, freedom of thought and expression and equality before the law. The issue has been considered by the Institute for the Future and by the U.K. Department of Trade and Industry.

Experts disagree whether specific and detailed laws will be required soon or safely in the distant future. Glenn McGee reports that sufficiently humanoid robots may appear by 2020. Ray Kurzweil sets the date at 2029. Another group of scientists meeting in 2007 supposed that at least 50 years had to pass before any sufficiently advanced system would exist.

The rules for the 2003 Loebner Prize competition envisioned possibility of robots having rights of their own:

61. If, in any given year, a publicly available open source Entry entered by the University of Surrey or the Cambridge Center wins the Silver Medal or the Gold Medal, then the Medal and the Cash Award will be awarded to the body responsible for the development of that Entry. If no such body can be identified, or if there is disagreement among two or more claimants, the Medal and the Cash Award will be held in trust until such time as the Entry may legally possess, either in the United States of America or in the venue of the contest, the Cash Award and Gold Medal in its own right.

Threat to Privacy

Aleksandr Solzhenitsyn's The First Circle describes the use of speech recognition technology in the service of tyranny. If an AI program exists that can understand natural languages and speech (e.g. English), then, with adequate processing power it could theoretically listen to every phone conversation and read every email in the world, understand them and report back to the program's operators exactly what is said and exactly who is saying it. An AI program like this could allow governments or other entities to efficiently suppress dissent and attack their enemies.

Threat to Human Dignity

Joseph Weizenbaum argued in 1976 that AI technology should not be used to replace people in positions that require respect and care, such as any of these:

- A customer service representative (AI technology is already used today for telephone-based interactive voice response systems)
- A therapist (as was seriously proposed by Kenneth Colby in the 1970s)
- A nursemaid for the elderly (as was reported by Pamela McCorduck in her book The Fifth Generation)
- A soldier
- A judge
- A police officer

Weizenbaum explains that we require authentic feelings of empathy from people in these positions. If machines replace them, we will find ourselves alienated, devalued and frustrated. Artificial intelligence, if used in this way, represents a threat to human dignity. Weizenbaum argues that the fact that we are entertaining the possibility of machines in these positions suggests that we have experienced an "atrophy of the human spirit that comes from thinking of ourselves as computers."

Pamela McCorduck counters that, speaking for women and minorities "I'd rather take my chances with an impartial computer," pointing out that there are conditions where we would prefer to have automated judges and police that have no personal agenda at all. AI founder John McCarthy objects to the moralizing tone of Weizenbaum's critique. "When moralizing is both vehement and vague, it invites authoritarian abuse," he writes.

Bill Hibbard writes that "Human dignity requires that we strive to remove our ignorance of the nature of existence, and AI is necessary for that striving."

Transparency and Open Source

Bill Hibbard argues that because AI will have such a profound effect on humanity, AI developers are representatives of future humanity and thus have an ethical obligation to be transparent in their efforts. Ben Goertzel and David Hart created OpenCog as an open source framework for AI development. OpenAI is a non-profit AI research company created by Elon Musk, Sam Altman and others to develop open source AI beneficial to humanity. There are numerous other open source AI developments.

Weaponization of Artificial Intelligence

Some experts and academics have questioned the use of robots for military combat, especially when such robots are given some degree of autonomous functions. The US Navy has funded a report which indicates that as military robots become more complex, there should be greater attention to implications of their ability to make autonomous decisions. One researcher states that autonomous robots might be more humane, as they could make decisions more effectively.

Within this last decade, there has been intensive research in autonomous power with the ability to learn using assigned moral responsibilities. "The results may be used when designing future military robots, to control unwanted tendencies to assign responsibility to the robots." From a consequentialist view, there is a chance that robots will develop the ability to make their own logical decisions on who to kill and that is why there should be a set moral framework that the A.I cannot override.

There has been a recent outcry with regards to the engineering of artificial-intelligence weapons and has even fostered up ideas of a robot takeover of mankind. AI weapons do present a type of danger different from that of human controlled weapons. Many governments have begun to fund programs to develop AI weaponry. The United States Navy recently announced plans to develop autonomous drone weapons, paralleling similar announcements by Russia and Korea respectively. Due to the potential of AI weapons becoming more dangerous than human operated weapons, Stephen Hawking and Max Tegmark have signed a Future of Life petition to ban AI weapons. The message posted by Hawking and Tegmark states that AI weapons pose an immediate danger and that action is required to avoid catastrophic disasters in the near future. "If any major military power pushes ahead with the AI weapon development, a global arms race is virtually inevitable, and the endpoint of this technological trajectory is obvious: autonomous weapons will become the Kalashnikov's of tomorrow", says the petition, which includes Skype co-founder Jaan Tallinn and MIT professor of linguistics Noam Chomsky as additional supporters against AI weaponry.

Physicist and Astronomer Royal Sir Martin Rees warned of catastrophic instances like "dumb robots going rogue or a network that develops a mind of its own." Huw Price, a colleague of Rees at Cambridge, has voiced a similar warning that humans may not survive when intelligence "escapes the constraints of biology." These two professors created the Centre for the Study of Existential Risk at Cambridge University in the hopes of avoiding this threat to human existence.

Regarding the potential for smarter-than-human systems to be employed militarily, the Open Philanthropy Project writes that this scenario "seem potentially as important as the risks related to loss of control", but that research organizations investigating AI's long-run social impact have spent relatively little time on this concern: "this class of scenarios has not been a major focus for the organizations that have been most active in this space, such as the Machine Intelligence Research Institute (MIRI) and the Future of Humanity Institute (FHI), and there seems to have been less analysis and debate regarding them".

Machine Ethics

Machine ethics (or machine morality) is the field of research concerned with designing Artificial Moral Agents (AMAs), robots or artificially intelligent computers that behave morally or as though moral.

Isaac Asimov considered the issue in the 1950s in his I, Robot. At the insistence of his editor John W. Campbell Jr., he proposed the Three Laws of Robotics to govern artificially intelligent systems. Much of his work was then spent testing the boundaries of his three laws to see where they would break down, or where they would create paradoxical or unanticipated behavior. His work suggests that no set of fixed laws can sufficiently anticipate all possible circumstances.

In 2009, during an experiment at the Laboratory of Intelligent Systems in the Ecole Polytechnique Fédérale of Lausanne in Switzerland, robots that were programmed to cooperate with each other (in searching out a beneficial resource and avoiding a poisonous one) eventually learned to lie to each other in an attempt to hoard the beneficial resource. One problem in this case may have been that the goals were "terminal" (i.e. in contrast, ultimate human motives typically have a quality of requiring never-ending learning).

Some experts and academics have questioned the use of robots for military combat, especially when such robots are given some degree of autonomous functions. The US Navy has funded a report which indicates that as military robots become more complex, there should be greater attention to implications of their ability to make autonomous decisions. The President of the Association for the Advancement of Artificial Intelligence has commissioned a study to look at this issue. They point to programs like the Language Acquisition Device which can emulate human interaction.

Vernor Vinge has suggested that a moment may come when some computers are smarter than humans. He calls this "the Singularity." He suggests that it may be somewhat or possibly very dangerous for humans. This is discussed by a philosophy called Singularitarianism.

The Machine Intelligence Research Institute has suggested a need to build "Friendly AI", meaning that the advances which are already occurring with AI should also include an effort to make AI intrinsically friendly and humane.

In 2009, academics and technical experts attended a conference organized by the Association for the Advancement of Artificial Intelligence to discuss the potential impact of robots and computers and the impact of the hypothetical possibility that they could become self-sufficient and able to make their own decisions. They discussed the possibility and the extent to which computers and robots might be able to acquire any level of autonomy, and to what degree they could use such abilities to possibly pose any threat or hazard. They noted that some machines have acquired various forms of semi-autonomy, including being able to find power sources on their own and being able to independently choose targets to attack with weapons. They also noted that some computer viruses can evade elimination and have achieved "cockroach intelligence." They noted that self-awareness as depicted in science-fiction is probably unlikely, but that there were other potential hazards and pitfalls.

In Moral Machines: Teaching Robots Right from Wrong, Wendell Wallach and Colin Allen conclude that attempts to teach robots right from wrong will likely advance understanding of human ethics by motivating humans to address gaps in modern normative theory and by providing a platform for experimental investigation. As one example, it has introduced normative ethicists to the controversial issue of which specific learning algorithms to use in machines. Nick Bostrom and Eliezer Yudkowsky have argued for decision trees (such as ID3) over neural networks and genetic algorithms on the grounds that decision trees obey modern social norms of transparency and predictability (e.g. stare decisis), while Chris Santos-Lang argued in the opposite direction on the grounds that the norms of any age must be allowed to change and that natural failure to fully satisfy these particular norms has been essential in making humans less vulnerable to criminal "hackers".

Unintended Consequences

Many researchers have argued that, by way of an "intelligence explosion" sometime in the 21st century, a self-improving AI could become so vastly more powerful than humans that we would not be able to stop it from achieving its goals. In his paper Ethical Issues in Advanced Artificial Intelligence, the Oxford philosopher Nick Bostrom even argues that artificial intelligence has the capability to bring about human extinction. He claims that general super-intelligence would be capable of independent initiative and of making its own plans, and may therefore be more appropriately thought of as an autonomous agent. Since artificial intellects need not share our human motivational tendencies, it would be up to the designers of the super-intelligence to specify its original motivations. In theory, a super-intelligent AI would be able to bring about almost any possible outcome and to thwart any attempt to prevent the implementation of its top goal, many uncontrolled unintended consequences could arise. It could kill off all other agents, persuade them to change their behavior, or block their attempts at interference.

However, the sheer complexity of human value systems makes it very difficult to make AI's motivations human-friendly. Unless moral philosophy provides us with a flawless ethical theory, an AI's utility function could allow for many potentially harmful scenarios that conform with a given ethical framework but not "common sense". According to Eliezer Yudkowsky, there is little reason to suppose that an artificially designed mind would have such an adaptation.

Bill Hibbard proposes an AI design that avoids several types of unintended AI behavior including self-delusion, unintended instrumental actions, and corruption of the reward generator.

Instead of overwhelming the human race and leading to our destruction Nick Bostrom believes that Super Intelligence can help us solve many difficult problems such as disease, poverty, and environmental destruction, and could help us to "enhance" ourselves.

Ethics of Artificial Intelligence in Fiction

The movie The Thirteenth Floor suggests a future where simulated worlds with sentient inhabitants are created by computer game consoles for the purpose of entertainment. The movie The Matrix suggests a future where the dominant species on planet Earth are sentient machines and humanity is treated with utmost Speciesism. The short story "The Planck Dive" suggest a future where humanity has turned itself into software that can be duplicated and optimized and the relevant distinction between types of software is sentient and non-sentient. The same idea can be found in the Emergency Medical Hologram of Starship Voyager, which is an apparently sentient copy of a reduced subset of the consciousness of its creator, Dr. Zimmerman, who, for the best motives, has created the system to give medical assistance in case of emergencies. The movies Bicentennial Man and A.I. deal with the possibility of sentient robots that could love. I, Robot explored some aspects of Asimov's three laws. All these scenarios try to foresee possibly unethical consequences of the creation of sentient computers.

The ethics of artificial intelligence is one of several core themes in BioWare's Mass Effect series of games. It explores the scenario of a civilization accidentally creating AI through a rapid increase in computational power through a global scale neural network. This event caused an ethical schism between those who felt bestowing organic rights upon the newly sentient Geth was appropriate and those who continued to see them as disposable machinery and fought to destroy them. Beyond the initial conflict, the complexity of the relationship between the machines and their creators is another ongoing theme throughout the story.

Over time, debates have tended to focus less and less on possibility and more on desirability, as emphasized in the "Cosmist" and "Terran" debates initiated by Hugo de Garis and Kevin Warwick. A Cosmist, according to Hugo de Garis, is actually seeking to build more intelligent successors to the human species.

Literature

The standard bibliography on ethics of AI is on PhilPapers. A recent collection is V.C. Müller(ed.) (2016).

MACHINE ETHICS

Machine ethics (or machine morality, computational morality, or computational ethics) is a part of the ethics of artificial intelligence concerned with the moral behavior of artificially intelligent beings. Machine ethics contrasts with roboethics, which is concerned with the moral behavior of humans as they design, construct, use and treat such beings. Machine ethics should not be confused with computer ethics, which focuses on professional behavior towards computers and information.

Before the 21st century the ethics of machines had largely been the subject of science fiction literature, mainly due to computing and artificial intelligence (AI) limitations. These limitations are being overcome through advances in theory and hardware resulting in a renewed focus on the field of artificial intelligence making Machine Ethics a bona fide field of research. The first use of the term seems to be by Mitchell Waldrop in the 1987 AI Magazine article "A Question of Responsibility":

"However, one thing that is apparent from the above discussion is that intelligent machines will embody values, assumptions, and purposes, whether their programmers consciously intend them to or not. Thus, as computers and robots become more and more intelligent, it becomes imperative that we think carefully and explicitly about what those built-in values are. Perhaps what we need is, in fact, a theory and practice of machine ethics, in the spirit of Asimov's three laws of robotics."

The field was delineated in the AAAI Fall 2005 Symposium on Machine Ethics:

"Past research concerning the relationship between technology and ethics has largely focused on responsible and irresponsible use of technology by human beings, with a few people being interested in how human beings ought to treat machines. In all cases, only human beings have engaged in ethical reasoning. The time has come for adding an ethical dimension to at least some machines. Recognition of the ethical ramifications of behavior involving machines, as well as recent and potential developments in machine autonomy, necessitate this. In contrast to computer hacking, software property issues, privacy issues and other topics normally ascribed to computer ethics, machine ethics is concerned with the behavior of machines towards human users and other machines. Research in machine ethics is key to alleviating concerns with autonomous systems—it could be argued that the notion of autonomous machines without such a dimension is at the root of all fear concerning machine intelligence. Further, investigation of machine ethics could enable the discovery of problems with current ethical theories, advancing our thinking about Ethics."

Machine ethics is sometimes referred to as machine morality, computational ethics or computational morality. A variety of perspectives of this nascent field can be found in the collected edition "Machine Ethics" that stems from the AAAI Fall 2005 Symposium on Machine Ethics.

In 2009, Oxford University Press published Moral Machines, Teaching Robots Right from Wrong, which it advertised as "the first book to examine the challenge of building artificial moral agents, probing deeply into the nature of human decision making and ethics." It cited some 450 sources, about 100 of which addressed major questions of machine ethics. Few sources were written before the 21st century largely because any form of artificial intelligence was nonexistent.

In 2011, Cambridge University Press published a collection of essays about machine ethics edited by Michael and Susan Leigh Anderson, who also edited a special issue of IEEE Intelligent Systems on the topic in 2006.

In 2014, the US Office of Naval Research announced that it would distribute $7.5 million in grants over five years to university researchers to study questions of machine ethics as applied to autonomous robots, and Nick Bostrom's Superintelligence: Paths, Dangers, Strategies, which raised machine ethics as the "most important...issue humanity has ever faced," reached #17 on the New York Times list of best selling science books.

Definitions

Moor distinguished between:

- Ethical impact agents, defined as machine systems carrying an ethical impact.
- Implicit ethical agents, with actions constrained to avoid unethical outcomes.
- Explicit ethical agents, which have algorithms to be consciously ethical.
- Full ethical agents, allowing that agents of the above three types derive their ethics from their human developers. (See artificial systems and moral responsibility.)

Foci

Urgency

In 2009, academics and technical experts attended a conference to discuss the potential impact of robots and computers and the impact of the hypothetical possibility that they could become self-sufficient and able to make their own decisions. They discussed the possibility and the extent to which computers and robots might be able to acquire any level of autonomy, and to what degree they could use such abilities to possibly pose any threat or hazard. They noted that some machines have acquired various forms of semi-autonomy, including being able to find power sources on their own and being able to independently choose targets to attack with weapons. They also noted that some computer viruses can evade elimination and have achieved "cockroach intelligence." They noted that self-awareness as depicted in science-fiction is probably unlikely, but that there were other potential hazards and pitfalls.

Some experts and academics have questioned the use of robots for military combat, especially when such robots are given some degree of autonomous functions. The US Navy has funded a report which indicates that as military robots become more complex, there should be greater attention to implications of their ability to make autonomous decisions. The President of the Association for the Advancement of Artificial Intelligence has commissioned a study to look at this issue. They point to programs like the Language Acquisition Device which can emulate human interaction.

Algorithms

Nick Bostrom and Eliezer Yudkowsky have argued for decision trees (such as ID3) over neural networks and genetic algorithms on the grounds that decision trees obey modern social norms of transparency and predictability (e.g. stare decisis), while Chris Santos-Lang argued in the opposite direction on the grounds that the norms of any age must be allowed to change and that natural failure to fully satisfy these particular norms has been essential in making humans less vulnerable than machines to criminal "hackers".

Training and Instruction

In 2009, during an experiment at the Laboratory of Intelligent Systems in the Ecole Polytechnique Fédérale of Lausanne in Switzerland, robots that were programmed to cooperate with each other in searching out a beneficial resource and avoiding a poisonous one eventually learned to lie to each other in an attempt to hoard the beneficial resource.

One problem in this case may have been that the goals were "terminal" (i.e. in contrast, ultimate human motives typically have a quality of requiring never-ending learning). The robots were grouped into clans and the successful members' digital genetic code was used for the next generation. After 50 successive generations, one clan's members discovered how to indicate food instead of poison. At the same time there were selfless robots who signaled danger and died to save others.These adaptations could have been the result of the genetic code or their human creators.(See also Genetic algorithm.)

Ethical Implications

In Moral Machines: Teaching Robots Right from Wrong, Wendell Wallach and Colin Allen conclude that attempts to teach robots right from wrong will likely advance understanding of human ethics by motivating humans to address gaps in modern normative theory and by providing a platform for experimental investigation.

Approaches

Several attempts have been made to make ethics computable or at least formal. Whereas Isaac Asimov's Three Laws of Robotics are usually not considered to be suitable for an artificial moral agent, it has been studied whether Kant's categorical imperative can be used. However, it has been pointed out that human value is in some aspects very complex. A way to explicitly surmount this difficulty is to receive human values directly from them through some mechanism, for example by learning them.

Another approach is to base current ethical considerations on previous similar situations. This is called casuistry and it could be implemented through research on the internet. The consensus from a million past decisions would lead to a new decision that is democracy dependent. This could, however, lead to decisions that reflect biases and unethical behaviors exhibited in society. The negative effects of this approach can be seen in Microsoft's Tay (bot), where the chatterbot learned to repeat racist and sexually charged messages sent by Twitter users.

One thought experiment focuses on a Genie Golem with unlimited powers presenting itself to the reader. This Genie declares that it will return in 50 years and demands that it be provided with a definite set of morals that it will then immediately act upon. The purpose of this experiment is to initiate a discourse over how best to handle defining complete set of ethics that computers may understand.

In Fiction

Isaac Asimov considered the issue in the 1950s in I, Robot. At the insistence of his editor John W. Campbell Jr., he proposed the Three Laws of Robotics to govern artificially intelligent systems. Much of his work was then spent testing the boundaries of his three laws to see where they would break down, or where they would create paradoxical or unanticipated behavior. His work suggests that no set of fixed laws can sufficiently anticipate all possible circumstances.

REFERENCES

1. Asimov, Isaac (1950). I, Robot.
2. Isaac Asimov (1964). “Introduction”. The Rest of the Robots. Doubleday. ISBN 0-385-09041-2.
3. Gunn, James (July 1980). “On Variations on a Robot”. IASFM: 56–81.Reprinted in James Gunn. (1982). Isaac Asimov: The Foundations of Science Fiction. Oxford u.a.: Oxford Univ. Pr. ISBN 0-19-503060-5.
4. Asimov, Isaac (1979). In Memory Yet Green. Doubleday. p. 237. ISBN 0-380-75432-0.
5. Asimov (1979), pp.236–8
6. Asimov (1979), p. 263.
7. Asimov (1979), pp. 285–7.
8. Asimov, Isaac (1979). In Memory Yet Green. Doubleday. Chapters 21 through 26 ISBN 0-380-75432-0.
9. Patrouch, Joseph F. (1974). The Science Fiction of Isaac Asimov. Doubleday. p. 42. ISBN 0-385-08696-2.
10. Asimov (1979), p. 620.
11. Asimov, Isaac (1980). In Joy Still Felt. Doubleday. p. 61. ISBN 0-385-15544-1.
12. Asimov, Isaac (November 1981). “The Three Laws”. Compute!. p. 18. Retrieved 26 October 2013.
13. Asimov, Isaac (12 April 2001). Robot Visions. Gollancz. ISBN 978-1-85798-336-4.
14. Gunn (1982).
15. “Isaac Asimov”. BBC. Retrieved 11 November 2010.
16. “Sci-fi writer Isaac Asimov”. Campus Star. The Daily Star. 29 July 2007.

Retrieved 7 August 2016. Only highly advanced robots (such as Daneel and Giskard) could comprehend this law.

17. Asimov, Isaac (1952). The Caves of Steel. Doubleday., translated by Jacques Brécard as Les Cavernes d'acier. J'ai Lu Science-fiction. 1975. ISBN 2-290-31902-3.
18. Patrouch (1974), p. 50.
19. Gunn (1980); reprinted in Gunn (1982), p. 69.
20. Jenkins, John H. (2002). "Review of "Cal"". Jenkins' Spoiler-Laden Guide to Isaac Asimov. Retrieved 2009-06-26.
21. Asimov, Isaac (1986). Robot Dreams (PDF). Retrieved 11 November 2010. "But you quote it in incomplete fashion. The Third Law is 'A robot must protect its own existence as long as such protection does not conflict with the First or Second Law.' " "Yes, Dr. Calvin. That is the Third Law in reality, but in my dream, the Law ended with the word 'existence'. There was no mention of the First or Second Law."
22. "'The Complete Robot' by Isaac Asimov". BBC. 3 November 2000. Retrieved 11 November 2010. The answer is that it had had its First Law modified
23. Asimov (1979), pp. 291–2.
24. Don D'Ammassa (2005). "Allen, Roger MacBride". Encyclopedia of science fiction. Infobase Publishing. p. 7. ISBN 978-0-8160-5924-9. Retrieved 11 November 2010.
25. "The Humanoids". Umich.edu. Retrieved 2015-03-28.
26. Heward Wilkinson (2009). The muse as therapist: a new poetic paradigm for psychotherapy. Karnac Books. pp. 22–23. ISBN 978-1-85575-595-6. Retrieved 11 November 2010.
27. MARK W. TIEDEMANN. Isaac Asimov's Aurora (ebook). Byron Press Visual Publications. p. 558. In short," Bogard said, "not all people are human
28. "Interview with Mark Tiedemann". Science Fiction and Fantasy World. 16 August 2002. Retrieved 2006-06-12.
29. Dilov, Lyuben (aka Lyubin, Luben or Liuben) (2002). Пътят на Икар. Захари Стоянов. ISBN 954-739-338-3.
30. Кесаровски, Никола (1983). Петият закон. Отечество.
31. Lawful Little Country: The Bulgarian Laws of Robotics | The Portal
32. Harrison, Harry (1989). The Fourth Law of Robotics. pp. 7–8. A robot must reproduce. As long as such reproduction does not interfere with the First or Second or Third Law
33. Ashrafian, Hutan (2014). "AIonAI: A Humanitarian Law of Artificial Intelligence and Robotics". Science and Engineering Ethics. Springer. 21 (1): 29–40. doi:10.1007/s11948-013-9513-9. PMID 24414678. Retrieved 20 January 2014.
34. Asimov, Isaac (1956–1957). The Naked Sun (ebook). p. 233. ... one robot poison an arrow without knowing it was using poison, and having a second robot hand

the poisoned arrow to the boy ...

35. Asimov, Isaac (1956–1957). The Naked Sun (ebook). p. 240. But a spaceship that was equipped with its own positronic brain would cheerfully attack any ship it was directed to attack, it seems to me. It would naturally assume all other ships were unmanned
36. Branislav L. Slantchev. "Foundation and Earth (1986)". gotterdammerung.org. Retrieved 11 November 2010.
37. Asimov, Isaac. Robots and Empire. Doubleday books. p. 151. ISBN 0-385-19092-1. although the woman looked as human as Daneel did, she was just as nonhuman
38. Butler, Gwendoline (2001). A Coffin for the Canary. Black Dagger Crime. ISBN 0-7540-8580-5.
39. Gunn (1980); reprinted in Gunn (1982), p. 73.
40. Asimov, Isaac (1982). "... That Thou Art Mindful Of Him". The Complete Robot. Nightfall, Inc. p. 611.
41. Gunn (1982), pp. 77–8.
42. Asimov, Isaac (1982). "The Bicentennial Man". The Complete Robot. Nightfall, Inc. p. 658.
43. Isaac Asimov. I, Robot (Asimov, Isaac - I, Robot.pdf). p. 122. Retrieved 11 November 2010.
44. Isaac Asimov. I, Robot (Asimov, Isaac - I, Robot.pdf). p. 75. Retrieved 11 November 2010.
45. Asimov, Isaac (1956–1957). The Naked Sun (ebook). p. 56. Are you trying to tell me, Daneel, that it hurts the robot to have me do its work? ... experience which the robot undergoes is as upsetting to it as pain is to a human
46. Murphy, Robin; Woods, David D. (July 2009). "Beyond Asimov: The Three Laws of Responsible Robotics" (PDF). IEEE Intelligent systems. 24 (4): 14–20. doi:10.1109/mis.2009.69. Retrieved July 2009. Check date values in: |access-date= (help)
47. Moravec, Hans. "The Age of Robots", Extro 1, Proceedings of the First Extropy Institute Conference on TransHumanist Thought (1994) pp. 84–100. June 1993 version available online.
48. "Rules for the modern robot". New Scientist (2544): 27. 27 March 2006. Retrieved 2006-06-12.
49. Sawyer, Robert J. (16 November 2007). "Guest Editorial: Robot Ethics". Science. 318 (5853): 1037. doi:10.1126/science.1151606. PMID 18006710. Retrieved 2010-10-10.
50. Sawyer, Robert J. (1991). "On Asimov's Three Laws of Robotics". Retrieved 2006-06-12.
51. Clarke, Roger. Asimov's laws of robotics: Implications for information technology. Part 1: IEEE Computer, December 1993, p53–61. Part 2: IEEE Computer, Jan

1994, p57–66. Both parts are available without fee at . Under "Enhancements to codes of ethics".

52. "Robotic age poses ethical dilemma". BBC News. 2007-03-07. Retrieved 2007-03-07.
53. Brin, David (1999). Foundation's Triumph. HarperCollins. ISBN 978-0-06-105241-5.
54. "Want Responsible Robotics? Start With Responsible Humans". Researchnews.osu.edu. Retrieved 2015-03-28.
55. Alan Winfield (2013-10-30). "Alan Winfield's Web Log: Ethical Robots: some technical and ethical challenges". Alanwinfield.blogspot.co.uk. Retrieved 2015-03-28.
56. "Principles of robotics - EPSRC website". Epsrc.ac.uk. Retrieved 2015-03-28.
57. Asimov, Isaac; Stanley Asimov (1995). Yours, Isaac Asimov: A Life in Letters. Doubleday. ISBN 0-385-47622-1.
58. Ellison, Harlan (1994). I, Robot: The illustrated screenplay. Aspect. ISBN 0-446-67062-6.
59. "Aliens (1986) - Memorable quotes". IMDb.com. Retrieved 2015-03-28.
60. "ALIEN". Angelfire.com. Retrieved 2015-03-28.
61. M. Keith Booker (2006). Alternate Americas: science fiction film and American culture. Greenwood Publishing Group. p. 205. ISBN 978-0-275-98395-6. Retrieved 11 November 2010. In a variation on Isaac Asimov's famous Three Laws of Robotics ...
62. "I, Rowboat". The Onion 29 July 1998.
63. Wagner, Roland Charles. "Three Laws of Robotic Sexuality".
64. https://www.linkedin.com/pulse/20140821045908-19801712-automota-why-robot-laws-will-never-be-effective
65. Stevens, Richard. Diesel Sweeties, strip 688.
66. Cham, Jorge. Piled Higher and Deeper, strip 440, 9 April 2004.
67. Frazer, J.D. "Illiad" (2007-03-21). "User Friendly". Retrieved 2007-03-21.
68. Stanley, Mark (2010-08-27). "Freefall". Retrieved 2010-08-27.
69. Veruggio, Gianmarco (2007). "The Roboethics Roadmap" (PDF). Scuola di Robotica: 2. Retrieved 28 March 2011.
70. Evans, Woody (2015). "Posthuman Rights: Dimensions of Transhuman Worlds". Universidad Complutense Madrid. Retrieved 7 April 2016.
71. The American Heritage Dictionary of the English Language, Fourth Edition
72. "Robots could demand legal rights". BBC News. December 21, 2006. Retrieved January 3, 2010.
73. Henderson, Mark (April 24, 2007). "Human rights for robots? We're getting carried away". The Times Online. The Times of London. Retrieved May 2, 2010.
74. McGee, Glenn. "A Robot Code of Ethics". The Scientist.

75. Kurzweil, Ray (2005). The Singularity is Near. Penguin Books. ISBN 0-670-03384-7.
76. The Big Question: Should the human race be worried by the rise of robots?, Independent Newspaper,
77. Loebner Prize Contest Official Rules — Version 2.0 The competition was directed by David Hamill and the rules were developed by members of the Robitron Yahoo group.
78. (McCorduck 2004, p. 308)
79. Joseph Weizenbaum, quoted in McCorduck 2004, pp. 356, 374–376
80. Hibbard, Bill (2014): "Ethical Artificial Intelligence".
81. Open Source AI. Bill Hibbard. 2008 proceedings of the First Conference on Artificial General Intelligence, eds. Pei Wang, Ben Goertzel and Stan Franklin.
82. OpenCog: A Software Framework for Integrative Artificial General Intelligence. David Hart and Ben Goertzel. 2008 proceedings of the First Conference on Artificial General Intelligence, eds. Pei Wang, Ben Goertzel and Stan Franklin.
83. Inside OpenAI, Elon Musk's Wild Plan to Set Artificial Intelligence Free Cade Metz, Wired 27 April 2016.
84. Call for debate on killer robots, By Jason Palmer, Science and technology reporter, BBC News, 8/3/09.
85. Robot Three-Way Portends Autonomous Future, By David Axe wired.com, August 13, 2009.
86. New Navy-funded Report Warns of War Robots Going "Terminator", by Jason Mick (Blog), dailytech.com, February 17, 2009.
87. Navy report warns of robot uprising, suggests a strong moral compass, by Joseph L. Flatley engadget.com, Feb 18th 2009.
88. http://search.proquest.com/docview/1372020233
89. Zach Musgrave and Bryan W. Roberts. "Why Artificial Intelligence Can Too Easily Be Weaponized - The Atlantic". The Atlantic.
90. Cat Zakrzewski. "Musk, Hawking Warn of Artificial Intelligence Weapons". WSJ.
91. GiveWell (2015). Potential risks from advanced artificial intelligence(Report). Retrieved 11 October 2015.
92. Anderson. "Machine Ethics". Retrieved 27 June 2011.
93. Anderson, Michael; Anderson, Susan Leigh, eds. (July 2011). Machine Ethics. Cambridge University Press. ISBN 978-0-521-11235-2.
94. Anderson, Michael; Anderson, Susan Leigh, eds. (July–August 2006). "Special Issue on Machine Ethics". IEEE Intelligent Systems. 21 (4): 10–63. doi:10.1109/mis.2006.70. ISSN 1541-1672.
95. Anderson, Michael; Anderson, Susan Leigh (Winter 2007). "Machine Ethics: Creating an Ethical Intelligent Agent". AI Magazine. American Association for Artificial Intelligence. 28 (4): 15–26. ISSN 0738-4602.

96. Asimov, Isaac (2008). I, Robot. New York: Bantam. ISBN 0-553-38256-X.
97. Evolving Robots Learn To Lie To Each Other, Popular Science, August 18, 2009
98. Santos-Lang, Chris (2002). "Ethics for Artificial Intelligences".
99. Science New Navy-funded Report Warns of War Robots Going "Terminator", by Jason Mick (Blog), dailytech.com, February 17, 2009.
100. Navy report warns of robot uprising, suggests a strong moral compass, by Joseph L. Flatley engadget.com, Feb 18th 2009.
101. AAAI Presidential Panel on Long-Term AI Futures 2008-2009 Study, Association for the Advancement of Artificial Intelligence, Accessed 7/26/09.
102. Scientists Worry Machines May Outsmart Man By JOHN MARKOFF, NY Times, July 26, 2009.
103. The Coming Technological Singularity: How to Survive in the Post-Human Era, by Vernor Vinge, Department of Mathematical Sciences, San Diego State University, (c) 1993 by Vernor Vinge.
104. Article at Asimovlaws.com Archived May 24, 2012, at the Wayback Machine., July 2004, accessed 7/27/09.
105. Wallach, Wendell; Allen, Colin (November 2008). Moral Machines: Teaching Robots Right from Wrong. USA: Oxford University Press. ISBN 978-0-19-537404-9.
106. Bostrom, Nick; Yudkowsky, Eliezer (2011). "The Ethics of Artificial Intelligence" (PDF). Cambridge Handbook of Artificial Intelligence. Cambridge Press.
107. Muehlhauser, Luke, and Louie Helm. 2012. "Intelligence Explosion and Machine Ethics". In Singularity Hypotheses: A Scientific and Philosophical Assessment, edited by Amnon Eden, Johnny Søraker, James H. Moor, and Eric Steinhart. Berlin: Springer.
108. Bostrom, Nick. 2003. "Ethical Issues in Advanced Artificial Intelligence". In Cognitive, Emotive and Ethical Aspects of Decision Making in Humans and in Artificial Intelligence, edited by Iva Smit and George E. Lasker, 12–17. Vol. 2. Windsor, ON: International Institute for Advanced Studies in Systems Research / Cybernetics.
109. Yudkowsky, Eliezer. 2011. "Complex Value Systems in Friendly AI". In Schmidhuber, Thórisson, and Looks 2011, 388–393.
110. "Sure, Artificial Intelligence May End Our World, But That Is Not the Main Problem". WIRED. Retrieved 2015-11-04.
111. Müller, Vincent C. (2016). "Risks of artificial intelligence". CRC Press - Chapman & Hall.
112. Moor, James H. (July–August 2006). "The Nature, Importance and Difficulty of Machine Ethics". IEEE Intelligent Systems. 21 (4): 18–21. doi:10.1109/MIS.2006.80.
113. Waldrop, Mitchell (Spring 1987). "A Question of Responsibility". AI Magazine. 8 (1): 28–39. doi:10.1609/aimag.v8i1.572.

114. "Papers from the 2005 AAAI Fall Symposium".
115. Anderson, Michael; Anderson, Susan Leigh, eds. (July 2011). Machine Ethics. Cambridge University Press. ISBN 978-0-521-11235-2.
116. Wallach, Wendell; Allen, Colin (November 2008). Moral Machines: Teaching Robots Right from Wrong. USA: Oxford University Press. ISBN 978-0-19-537404-9.
117. Anderson, Michael; Anderson, Susan Leigh, eds. (July–August 2006). "Special Issue on Machine Ethics". IEEE Intelligent Systems. 21 (4): 10–63. doi:10.1109/mis.2006.70. ISSN 1541-1672.
118. Tucker, Patrick (13 May 2014). "Now The Military Is Going To Build Robots That Have Morals". Defense One. Retrieved 9 July 2014.
119. "Best Selling Science Books". New York Times. New York Times. September 8, 2014. Retrieved 9 November 2014.
120. Scientists Worry Machines May Outsmart Man By JOHN MARKOFF, NY Times, July 26, 2009.
121. Call for debate on killer robots, By Jason Palmer, Science and technology reporter, BBC News, 8/3/09.
122. Science New Navy-funded Report Warns of War Robots Going "Terminator", by Jason Mick (Blog), dailytech.com, February 17, 2009.
123. Navy report warns of robot uprising, suggests a strong moral compass, by Joseph L. Flatley engadget.com, Feb 18th 2009.
124. AAAI Presidential Panel on Long-Term AI Futures 2008-2009 Study, Association for the Advancement of Artificial Intelligence, Accessed 7/26/09.
125. Bostrom, Nick; Yudkowsky, Eliezer (2011). "The Ethics of Artificial Intelligence" (PDF). Cambridge Handbook of Artificial Intelligence. Cambridge Press.
126. Santos-Lang, Chris (2002). "Ethics for Artificial Intelligences".
127. Santos-Lang, Christopher (2014). "Moral Ecology Approaches to Machine Ethics". In van Rysewyk, Simon; Pontier, Matthijs. Machine Medical Ethics(PDF). Switzerland: Springer. pp. 111–127. doi:10.1007/978-3-319-08108-3_8.
128. Evolving Robots Learn To Lie To Each Other, Popular Science, August 18, 2009
129. Anderson, Susan Leigh (2011): The Unacceptability of Asimov's Three Laws of Robotics as a Basis for Machine Ethics. In: Machine Ethics, ed. Michael Anderson, Susan Leigh Anderson. New York: Oxford University Press. pp.285-296. ISBN 9780511978036
130. Powers, Thomas M. (2011): Prospects for a Kantian Machine. In: Machine Ethics, ed. Michael Anderson, Susan Leigh Anderson. New York: Oxford University Press. pp.464-475.
131. Muehlhauser, Luke, Helm, Louie (2012): Intelligence Explosion and Machine Ethics.
132. Yudkowsky, Eliezer (2004): Coherent Extrapolated Volition.
133. Guarini, Marcello (2011): Computational Neural Modeling and the Philosophy of

Ethics. Reflections on the Particularism-Generalism Debate. In: Machine Ethics, ed. Michael Anderson, Susan Leigh Anderson. New York: Oxford University Press. pp.316-334.

134. Hibbard, Bill (2014): Ethical Artificial Intelligence. http://arxiv.org/abs/1411.1373

135. Anderson, M. and Anderson, S. (2007). Creating an Ethical Intelligent Agent. AI Magazine, Volume 28(4).

136. "Microsoft chatbot is taught to swear on Twitter - BBC News". BBC News. Retrieved 2016-04-17.

137. Nazaretyan, A. (2014). A. H. Eden, J. H. Moor, J. H. Søraker and E. Steinhart (eds): Singularity Hypotheses: A Scientific and Philosophical Assessment. Minds & Machines, 24(2), pp.245-248.

138. Asimov, Isaac (2008). I, robot. New York: Bantam. ISBN 0-553-38256-X.

CHAPTER 6

Mobile Robots

Chapter Contents

A mobile robot is one which can move around its environment independently. In order to do this the robot needs to be able to navigate, and the range and accuracy of navigational ability required will vary depending on the size of the robot and the type of task it carries out. Types of mobile robot currently in production include the domestic robot, which comes in a variety of forms ranging from the robot vacuum cleaner to the robot dog. Potential future uses of the mobile robot include robot paramedics, farmhands and border guards.

INTRODUCTION

Mobile robots have the capability to move around in their environment and are not fixed to one physical location. Mobile robots can be "autonomous" (AMR - autonomous mobile robot) which means they are capable of navigating an uncontrolled environment without the need for physical or electro-mechanical guidance devices. Alternatively, mobile robots can rely on guidance devices that allow them to travel a pre-defined navigation route in relatively controlled space (AGV - autonomous guided vehicle). By contrast, industrial robots are usually more-or-less stationary, consisting of a jointed arm (multi-linked manipulator) and gripper assembly (or end effector), attached to a fixed surface. Mobile robots have become more commonplace in commercial and industrial settings. Hospitals have been using autonomous mobile robots to move materials for many years. Warehouses have installed mobile robotic systems to efficiently move materials from stocking shelves to order fulfillment zones. Mobile robots are also a major focus of current research and almost every major university has one or more labs that focus on mobile robot research. Mobile robots are also found in industrial, military and security settings. Domestic robots are consumer products, including entertainment robots and those that perform certain household tasks such as vacuuming or gardening.

The components of a mobile robot are a controller, control software, sensors and actuators. The controller is generally a microprocessor, embedded microcontroller or a personal computer (PC). Mobile control software can be either assembly level language or high-level languages such as C, C++, Pascal, Fortran or special real-time software. The sensors used are dependent upon the requirements of the robot. The requirements could be dead reckoning, tactile and proximity sensing, triangulation ranging, collision avoidance, position location and other specific applications.

A spying robot is an example of a mobile robot capable of movement in a given environment.

Classification

Mobile robots may be classified by:

- The environment in which they travel:

  Land or home robots are usually referred to as Unmanned Ground Vehicles (UGVs). They are most commonly wheeled or tracked, but also include legged robots with two or more legs (humanoid, or resembling animals or insects).

  - Delivery & Transportation robots can move materials and supplies through a work environment

- Aerial robots are usually referred to as Unmanned Aerial Vehicles (UAVs)
- Underwater robots are usually called autonomous underwater vehicles (AUVs)
- Polar robots, designed to navigate icy, crevasse filled environments

■ The device they use to move, mainly:
- Legged robot : human-like legs (i.e. an android) or animal-like legs.
- Wheeled robot.
- Tracks.

LEGGED ROBOT

Legged robots are a type of mobile robot. They are somewhat a recent innovation in robotics. However, many or all bipedal models are not practical because they are cumbersome and slow. Most successful legged robots have four or six legs for further stability. This legs-over-wheels approach lends itself for use in all-terrain purposes because legs are more effective in an uneven environment than wheels.

Legged robots include:

■ Toy robots such as the bipedal QRIO and ASIMO, and the quadruped AIBO
■ Military robots such as LS3
■ one-legged hopping robots sometimes called pogo stick robots,
■ bipedal robots, quadrupedal robots and hexapod robots.

A hexapod robot

CONTINUOUS TRACK

Continuous track, also called tank tread or caterpillar track, is a system of vehicle propulsion in which a continuous band of treads or track plates is driven by two or more wheels. This band is typically made of modular steel plates in the case of military vehicles and heavy equipment, or synthetic rubber reinforced with steel wires in the case of lighter agricultural or construction vehicles.

The large surface area of the tracks distributes the weight of the vehicle better than steel or rubber tyres on an equivalent vehicle, enabling a continuous tracked vehicle to traverse soft ground with less likelihood of becoming stuck due to sinking. The prominent treads of the metal plates are both hard-wearing and damage resistant, especially in comparison to rubber tyres. The aggressive treads of the tracks provide good traction in soft surfaces but can damage paved surfaces, so some metal tracks can have rubber pads installed for use on paved surfaces.

Continuous tracks can be traced back as far as 1770 and today are commonly used on a variety of vehicles including bulldozers, excavators, tanks, and tractors, but can be found on any vehicle used in an application that can benefit from the added traction, low ground pressure and durability inherent in continuous track propulsion systems.

A closeup of continuous tracks on a bulldozer

A British Army Challenger 1 tank

An agricultural tractor with rubber tracks

In modern times, continuous track propulsion systems can be traced back to a crude continuous track system designed in the 1770s by Richard Lovell Edgeworth. Polish mathematician and inventor Józef Maria Hoene-Wroński conceived of the idea in the 1830s. The British polymath Sir George Cayley patented a continuous track, which he called a universal railway. In 1837, a Russian inventor Dmitry Zagryazhsky designed a "carriage with mobile tracks" which he patented the same year, but due to a lack of funds and interest from manufacturers he was unable to build a working prototype, and his patent was voided in 1839.

Dreadnaught Wheel

Although not a continuous track in the form encountered today, a dreadnaught wheel or "endless railway wheel" was patented by the British Engineer James Boydell in 1846. In Boydell's design, a series of flat feet are attached to the periphery of the wheel, spreading the weight. A number of horse-drawn wagons, carts and gun carriages were successfully deployed in the Crimean War, waged between October 1853 and February 1856, the Royal Arsenal at Woolwich manufacturing dreadnaught wheels. A letter of recommendation was signed by Sir William Codrington, the General commanding the troops at Sebastapol.

Boydell patented improvements to his wheel in 1854 (No. 431) - the year his dreadnaught wheel was first applied to a steam engine - and 1858 (No. 356), the latter an impracticable palliative measure involving the lifting one or other of the driving wheels to facilitate turning.

A number of manufacturers including Richard Bach, Richard Garrett & Sons, Charles Burrell & Sons and Clayton & Shuttleworth applied the Boydell patent under licence. The British military were interested in Boydell's invention from an early date. One of the objectives was to transport Mallet's Mortar, a giant 36 weapon which was under development but, by the end of the Crimean war, the mortar was not ready for service. A detailed report of the tests on steam traction, carried out by a select Committee of the Board of Ordnance, was published in June 1856, by which date the Crimean War was over, consequently the mortar and its transportation became irrelevant. In those tests, a Garrett engine was put through its paces on Plumstead Common. The Garrett engine featured in the Lord Mayor's show in London, and in the following month that engine was shipped to Australia. A steam tractor employing dreadnaught wheels was built at Bach's Birmingham works, and was used between 1856 and 1858 for ploughing in Thetford; and the first generation of Burrell/Boydell engines was built at the St. Nicholas works in 1856, again, after the close of the Crimean war. Between late 1856 and 1862 Burrell manufactured not less than a score of engines fitted with dreadnaught wheels. And in April 1858, The Engineer gave a brief description of a Clayton & Shuttleworth engine fitted with dreadnaught wheels, which was supplied not to the Western Allies, but to the Russian Government for heavy artillery haulage in the Crimea, in the post-war period. Steam tractors fitted with dreadnaught wheels had a number of shortcomings and, notwithstanding the creations of the late 1850s, were never used extensively.

In August 1858, more than two years after the end of the Crimean War, John Fowler (agricultural engineer) filed British Patent No. 1948 on another form of Endless Railway. In his illustration of the invention, Fowler used a pair of wheels of equal diameter on each side of his vehicle, around which pair of toothed wheels ran a ‹track' of eight jointed segments, with a smaller jockey/drive wheel between each pair of wheels, to support the ‹track'. Comprising only eight sections, the ‹track' sections are essentially ‹longitudinal', as in Boydell's initial design. Fowler's arrangement is a precursor to the multi-section caterpillar track in which a relatively large number of short ‹transverse' treads are used, as proposed by Sir George Caley in 1825, rather than a small number of relatively long ‹longitudinal' treads.

Further to Fowler's patent of 1858, in 1877, a Russian, Fyodor Blinov, created a tracked vehicle called wagon moved on endless rails (caterpillars). It lacked self-propulsion and was pulled by horses. Blinov received a patent for his wagon in 1878. From 1881 to 1888 he developed a steam-powered caterpillar-tractor. This self-propelled crawler was successfully tested and featured at a farmers' exhibition in 1896.

Steam traction engines were used at the end of the 19th Century in the Boer Wars. But neither dreadnaught wheels nor continuous tracks were used, rather roll-out wooden plank roads were thrown under the wheels as required.

In short, whilst the development of the continuous track engaged the attention of a number of inventors in the 18th and 19th centuries, the general use and exploitation of the continuous track belonged to the 20th Century.

20th Century

A little-known American inventor, Henery T. Stith, developed a continuous track prototype which was, in multiple forms, patented in 1873, 1880, and 1900. The last was for the application of the track to a prototype off-road bicycle built for his son. The 1900 prototype is retained by his surviving family.

Frank Beamond, a less-commonly known but significant British inventor, designed and built caterpillar tracks, and was granted patents for them in a number of countries, in 1900 and 1907.

Lombard Steam Log Hauler (Designed, patented 1901)

Commercial Success

An effective continuous track was invented and implemented by Alvin Orlando Lombard for the Lombard Steam Log Hauler. He was granted a patent in 1901 and built the first steam-powered log hauler at the Waterville Iron Works in Waterville, Maine, the same year. In all, 83 Lombard steam log haulers are known to have been built up to 1917, when production switched entirely to internal combustion engine powered machines, ending with a Fairbanks diesel-powered unit in 1934. Undoubtedly, Alvin Lombard was the first commercial manufacturer of the tractor crawler.

At least one of Lombard's steam-powered machines apparently remains in working order. A gasoline-powered Lombard hauler is on display at the Maine State Museum in Augusta. In addition, there may have been up to twice as many Phoenix Centipeed versions of the steam log hauler built under license from Lombard, with vertical instead of horizontal cylinders. In 1903, the founder of Holt Manufacturing, Benjamin Holt, paid Lombard $60,000 for the right to produce vehicles under his patent.

Hornsby tracked, paraffin-powered tractor, c. 1908

Hornsby tractor, 1909

At about the same time a British agricultural company, Hornsby in Grantham, developed a continuous track which was patented in 1905. The design differed from modern tracks in that it flexed in only one direction, with the effect that the links locked together to form a solid rail on which the road wheels ran. Hornsby's tracked vehicles were given trials as artillery tractors by the British Army on several occasions between 1905 and 1910, but not adopted. The Hornsby tractors featured a track-steer clutch arrangement, which is the basis of the modern crawler operation. The patent was purchased by Holt.

Military Application

Continuous track was first applied to a military vehicle on the British prototype tank Little Willie. British Army officers, Colonel Ernest Swinton and Colonel Maurice Hankey, became convinced that it was possible to develop a fighting vehicle that could provide protection from machine gun fire.

Caterpillar

The name came from a soldier during the tests on the Hornsby crawler, '...trials began at Aldershot in July 1907. The soldiers immediately christened the 70bhp No.2 machine the "caterpillar"'.'

Holt adopted that name for his "crawler" tractors. Holt began moving from steam to gasoline-powered designs, and in 1908 brought out the 40 horsepower "Holt Model 40 Caterpillar". Holt incorporated the Holt Caterpillar Company, in early 1910, later that year trademarked the name "Caterpillar" for his continuous tracks.

In a memorandum of 1908, Antarctic explorer Robert Falcon Scott presented his view that man-hauling to the South Pole was impossible and that motor traction was needed. Snow vehicles did not yet exist however, and so his engineer Reginald Skelton developed the idea of a caterpillar track for snow surfaces. These tracked motors were built by the Wolseley Tool and Motor Car Company in Birmingham, tested in Switzerland and Norway, and can be seen in action in Herbert Ponting's 1911 documentary film of Scott's Antarctic Terra Nova Expedition (at minute 50, here). Scott died during the expedition in 1912, but expedition member and biographer Apsley Cherry-Garrard credited Scott's "motors" with the inspiration for the British World War I tanks, writing: *Scott never knew their true possibilities; for they were the direct ancestors of the 'tanks' in France*.

During World War I Holt tractors were used by the British and Austro-Hungarian armies to tow heavy artillery and stimulated the development of tanks in several countries. The first tanks to go into action, the Mark I, built by Great Britain, were designed from scratch and were inspired by, but not directly based on the Holt. The slightly later French and German tanks were built on modified Holt running gear.

A modern Caterpillar Inc. D9T bulldozer

Caterpillar Tractor Company began in 1925 from a merger of the Holt Manufacturing Company and the C. L. Best Tractor Company; an early successful manufacturer of crawler tractors.

With the Caterpillar D10 in 1977 Caterpillar resurrected a design by Holt and Best, the high-sprocket-drive, since known as the "High Drive", which had the advantage of keeping the main drive shaft away from ground shocks and dirt, and is still used in their larger dozers.

Patent History

A long line of patents disputes who was the "originator" of continuous tracks. There were a number of designs that attempted to achieve a track laying mechanism, although these designs do not generally resemble modern tracked vehicles.

Blinov

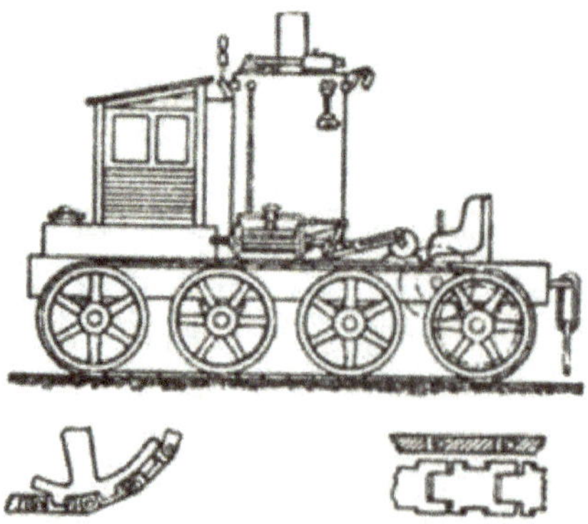

The draft of Blinov's steam-powered continuous track tractor

In 1877 Russian inventor Fyodor Abramovich Blinov created a tracked vehicle called "wagon moved on endless rails" (caterpillars). It lacked self-propelling and was horse-drawn. Blinov got a patent for his wagon the next year. Later, in 1881-1888 he created a steam-powered caterpillar-tractor. This self-propelled crawler was successfully tested and showed at a farmers' exhibition in 1896.

Dinsmoor

According to *Scientific American*, it was Charles Dinsmoor of Warren, Pennsylvania that invented a "vehicle" that was of endless tracks. The article gives a detailed description of the endless tracks and the illustration looks much like today's tracked vehicles. The invention has been patented as No. 351,749 on November 2, 1886.

Lombard

Alvin O. Lombard of Waterville, Maine was issued a patent in 1901 for the Lombard Steam Log Hauler that resembles a regular railroad steam locomotive with sled steerage on front and crawlers in rear for hauling logs in the Northeastern United States and Canada. The haulers allowed pulp to be taken to rivers in the winter. Prior to then, horses could be used only until snow depths made hauling impossible. Lombard began commercial production which lasted until around 1917 when focus switched entirely to gasoline powered machines. A gasoline-powered hauler is on display at the Maine State Museum in Augusta, Maine.

Hornsby/Holt/Phoenix

A Model of the Hornsby tractor

After Lombard began operations, Hornsby in England manufactured at least two full length "track steer" machines, and their patent was later purchased by Holt in 1913, allowing Holt to claim to be the "inventor" of the crawler tractor. Since the "tank" was a British concept it is more likely the Hornsby, which had been built and unsuccessfully pitched to their military, was the inspiration.

In a patent dispute involving rival crawler builder Best, testimony was brought in from people including Lombard, that Holt had inspected a Lombard log hauler shipped out to a western state by people who would later build the Phoenix log hauler in Eau Claire, Wisconsin, under license from Lombard. The Phoenix Centipeed typically had a fancier wood cab, steering wheel tipped forward at a 45 degree angle and vertical instead of horizontal cylinders.

Linn

In the meantime a gasoline-powered motor home was built by Lombard for Holman Harry (Flannery) Linn of Old Town, Maine to pull the equipment wagon of his dog & pony show, resembling a trolley car only with wheels in front and Lombard crawlers in rear. Linn had experimented with gasoline and steam-powered vehicles and six-wheel drive before this, and at some point entered Lombard's employment as a demonstrator, mechanic and sales agent. This resulted in a question of proprietorship of patent rights after a single rear-tracked gasoline-powered road engine of tricycle arrangement was built to replace the larger motor home in 1909 on account of problems with the old picturesque wooden bridges. This dispute resulted in Linn departing Maine and relocating to Morris, New York, to build an improved, contour following flexible lag tread or crawler with independent suspension of halftrack type, gasoline and later diesel powered. Although several were delivered for military use between 1917 and 1946, Linn never received any large military orders. Most of the production between 1917 and 1952, approximately 2500 units, was sold directly to highway departments and contractors. Steel tracks and payload capacity allowed these machines to work in terrain that would typically cause the poorer quality rubber tyres that existed before the mid-1930s to spin uselessly, or shred completely.

Linn was a pioneer in snow removal before the practice was embraced in rural areas, with a nine-foot steel v-plow and sixteen foot adjustable leveling wings on either side. Once the highway system became paved, snowplowing could be done by four wheel drive trucks equipped by improving tyre designs, and the Linn became an off highway vehicle, for logging, mining, dam construction, arctic exploration, etc.

Sloan

Once steel cleats became unpopular on paved roads, in 1938 a limited experiment began to bridge the gap between truck and tractor, a "convertible vehicle" patented by Phillip Sloan, the C5 Catruk, suffered design flaws and limited production before it was finally abandoned.

Engineering

Construction and Operation

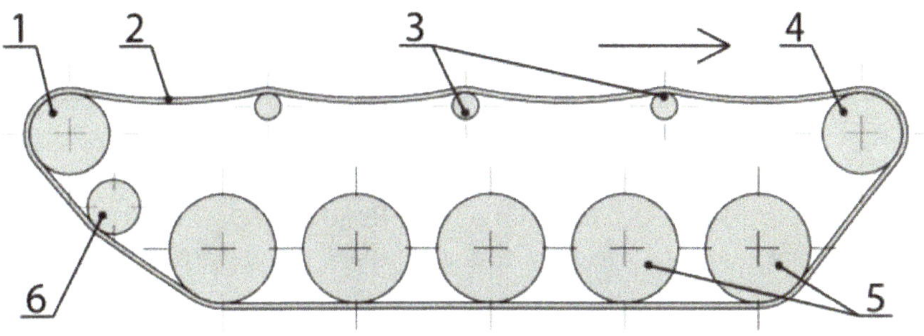

Diagram of tracked suspension.(1=rear drive wheel (rear wheel drive), 2=track, 3=return rollers, 4=front drive wheel (front wheel drive), 5=road wheels, 6=idler)

A sprocket wheel on a tank

Modern tracks are built from modular chain links which together compose a closed chain. The links are jointed by a hinge, which allows the track to be flexible and wrap around a set of wheels to make an endless loop. The chain links are often broad, and can be made of manganese alloy steel for high strength, hardness, and abrasion resistance.

Track construction and assembly is dictated by the application. Military vehicles use a track shoe that is integral to the structure of the chain in order to reduce track weight. Reduced weight allows the vehicle to move faster and decreases overall vehicle weight to ease transportation. Since track weight is completely unsprung, reducing it improves suspension performance at speeds where the track's momentum is significant. In contrast, agricultural and construction vehicles opt for a track with shoes that attach to the chain with bolts and do not form part of the chain's structure. This allows track shoes to break without compromising the ability of the vehicle to move and decrease productivity but increases the overall weight of the track and vehicle. Extra weight is an advantage when optimizing for traction and power over speed and mobility.

The vehicle's weight is transferred to the bottom length of track by a number of road wheels, or sets of wheels called bogies. Road wheels are typically mounted on some form of suspension to cushion the ride over rough ground. Suspension design in military vehicles is a major area of development; the very early designs were often completely unsprung. Later-developed road wheel suspension offered only a few inches of travel using springs, whereas modern hydro-pneumatic systems allow several feet of travel and include shock absorbers. Torsion-bar suspension has become the most common type of military vehicle suspension. Construction vehicles have smaller road wheels that are designed primarily to prevent track derailment and they are normally contained in a single bogie that includes the idler-wheel and sometimes the sprocket.

Transfer of power to the track is accomplished by a drive wheel, or *drive sprocket*, driven by the motor and engaging with holes in the track links or with pegs on them to drive the track. In military vehicles, the drive wheel is typically mounted well above the contact area on the ground, allowing it to be fixed in position. In agricultural crawlers it is normally incorporated as part of the bogie. Placing suspension on the sprocket is possible, but is mechanically more complicated. A non-powered wheel, an *idler*, is placed at the opposite end of the track, primarily to tension the track, since loose track could be easily thrown (slipped) off the wheels. To prevent throwing, the inner surface of the track links usually have vertical guide horns engaging grooves, or gaps between the doubled road and idler/sprocket wheels. In military vehicles with a rear sprocket, the idler wheel is placed higher than the road wheels to allow it to climb over obstacles. Some track arrangements use return rollers to keep the top of the track running straight between the drive sprocket and idler. Others, called *slack track*, allow the track to droop and run along the tops of large road wheels. This was a feature of the Christie suspension, leading to occasional misidentification of other slack track-equipped vehicles.

Overlapping Road Wheels

Overlapped and interleaved road wheels of a German Tiger I heavy tank

An Sd.Kfz. 11's half-track units, showing the rims of its six *Schachtellaufwerk* overlapped/ interleaved roadwheel sets for each track unit per side

Many World War II German military vehicles, initially (starting in the late 1930s) including all vehicles originally designed to be half-tracks and all later tank designs (after the Panzer IV), had slack-track systems, usually driven by a front-located drive sprocket, the track returning along the tops of a design of overlapping and sometimes interleaved large diameter road wheels, as on the suspension systems of the Tiger I and Panther tanks, generically known by the term *Schachtellaufwerk* in German, for both half-track and fully tracked vehicles. There were suspensions with one (sometimes double) wheel per axle, alternately supporting the inner and outer side of the track, and interleaved suspensions with two or three road wheels per axle, distributing the load over the track. The choice of overlapping/interleaved road wheels allowed the use of slightly more torsion bar suspension members, allowing any German tracked military vehicle with such a setup to have a noticeably smoother ride over challenging terrain, leading to reduced wear and more accurate fire. There were some major disadvantages with this though, one being on the Russian front, mud would get stuck in-between the overlapping wheels, and would freeze, immobilizing the vehicle. As a tracked vehicle moves, the load of each wheel moves over the track, pushing down and forward that part of the earth, snow, etc. under it, similarly to a wheeled vehicle but to a lesser extent because the tread helps distribute the load. Apparently, on some surfaces, this consumes enough energy to slow the vehicle down significantly, so overlapped and interleaved wheels improve performance (including fuel consumption) by loading the track more evenly. It also must have extended the life of the tracks and possibly of the wheels. The wheels also better protect the vehicle from enemy fire, and mobility is improved when some wheels are missing. But this complicated approach has not been used since World War II ended. This may be related more to maintenance than to original cost. The torsion bars and bearings may stay dry and clean, but the wheels and tread work in mud, sand, rocks, snow and so on. In addition, the outer wheels (up to 9 of them, some double) had to be removed to access the inner ones. In WW II, vehicles typically had to be maintained a few months before being destroyed or captured, but in peacetime vehicles must train several crews, over a period of decades.

Advantages

A Russian tracked vehicle designed to operate on snow and swamps

Tracked vehicles have better mobility over rough terrain than those with wheels. They smooth out the bumps, glide over small obstacles and are capable of crossing trenches or breaks in the terrain. Riding in a fast tracked vehicle feels like riding in a boat over heavy swells. Tracks cannot be punctured or torn. Tracks are much less likely to get stuck in soft ground, mud, or snow since they distribute the weight of the vehicle over a larger contact area, decreasing its ground pressure. In addition, the larger contact area, coupled with the cleats, or grousers, on the track shoes, allows vastly superior traction that results in a much better ability to push or pull large loads where wheeled vehicles would dig in. Bulldozers, which are most often tracked, use this attribute to rescue other vehicles, (such as wheel loaders) which have become stuck in, or sunk into, the ground. Tracks can also give higher maneuverability, as tracked vehicles can turn in place without forward or backward movement by driving the tracks in opposite directions. In addition, should a track be broken, assuming the correct tools are available, it can be repaired without the need for special facilities; something which is crucial in a combat situation.

The seventy-ton M1 Abrams tank has an average ground pressure of just over 15 psi (100 kPa). Since tyre air pressure is approximately equal to average ground pressure, a typical car will have an average ground pressure of 28 psi (190 kPa) to 33 psi (230 kPa).

Disadvantages

Tracked vehicles may be put on semitrailers or railway cars for long-distance hauling.

The disadvantages of tracks are lower top speed, much greater mechanical complexity, shorter life and the damage that their all-steel versions cause to the surface on which they pass. They are assumed to severely damage hard terrain such as asphalt pavement, but actually have significantly lower ground pressures than equivalent or lighter wheeled vehicles. However, they often cause damage to less firm terrain such as lawns, gravel roads, and farm fields, as the sharp edges of the track easily rout the turf. Accordingly, vehicle laws and local ordinances often require rubberised tracks or track pads. A compromise between all-steel and all-rubber tracks exists: attaching rubber pads to individual track links ensures that continuous track vehicles can travel more smoothly, quickly, and quietly on paved surfaces. While these pads slightly reduce a vehicle's cross-country traction, in theory they prevent damage to any pavement.

Small tracks on a roadworks machine, note the rubber pads to reduce wear on the carriageway.

Worn and new track pads on an M1 Abrams battle tank.

Additionally, the loss of a single segment in a track immobilizes the entire vehicle, which can be a disadvantage in situations where high reliability is important. Tracks can also ride off their guide wheels, idlers or sprockets, which can cause them to jam or to come completely off the guide system (this is called a 'thrown' track). Jammed tracks may become so tight that the track may need to be broken before a repair is possible, which requires either explosives or special tools. Multi-wheeled vehicles, for example, 8 X 8 military vehicles, may often continue driving even after the loss of one or more non-sequential wheels, depending on the base wheel pattern and drive train.

Many manufacturers provide rubber tracks instead of steel, especially for agricultural applications. Rather than a track made of linked steel plates, a reinforced rubber belt with chevron treads is used. In comparison to steel tracks, rubber tracks are lighter, make less noise, create less maximal ground pressure and do not damage paved roads. The disadvantage is that they are not as solid as steel tracks. Previous belt-like systems, such as those used for half-tracks in World War II, were not as strong, and during military actions were easily damaged. The first rubber track was invented and constructed by Adolphe Kégresse and patented in 1913; rubber tracks are often called Kégresse tracks.

Prolonged use places enormous strain on the drive transmission and the mechanics of the tracks, which must be overhauled or replaced regularly. It is common to see tracked vehicles such as bulldozers or tanks transported long distances by a wheeled carrier such as a tank transporter or train, though technological advances have made this practice less common among tracked military vehicles than it once was.

"Live" and "Dead" Track

Tracks may be broadly categorized as "live" or "dead" track. "Dead" track is a simple design in which each track plate is connected to the rest with hinge-type pins. These dead tracks will lie flat if placed on the ground; the drive sprocket pulls the track around the wheels with no assistance from the track itself. "Live" track is slightly more complex, with each link connected to the next by a bushing which causes the track to bend slightly inward. A length of live track left on the ground will curl upward slightly at each end. Although the drive sprocket must still pull the track around the wheels, the track itself tends to bend inward, slightly assisting the sprocket and somewhat conforming to the wheels.

Current Manufacturers

Caterpillar D9 tracks

The pioneer manufacturers have been replaced mostly by large tractor companies such as AGCO, Liebherr Group, John Deere, Yanmar, New Holland, Kubota, Case, Caterpillar Inc., CLAAS. Also, there are some crawler tractor companies specialising in niche markets. Examples are Otter Mfg. Co. and Struck Corporation., with many wheeled vehicle conversion kits available from the American Mattracks firm of Minnesota since the mid-1990s.

Russian off-road vehicles are built by companies such as ZZGT and Vityaz.

In Nature

- *Navicula* diatoms are known for their ability to creep about on each other and on hard surfaces such as microscope slides. It is thought that around the outside of the navicula's shell is a girdle of protoplasm that can flow and thus act as a tank track.

Mobile Robot Navigation

There are many types of mobile robot navigation:

Manual Remote Or Tele-Op

A manually teleoperated robot is totally under control of a driver with a joystick or other control device. The device may be plugged directly into the robot, may be a wireless joystick, or may be an accessory to a wireless computer or other controller. A tele-op'd robot is typically used to keep the operator out of harm's way. Examples of manual remote robots include Robotics Design's ANATROLLER ARI-100 and ARI-50, Foster-Miller's Talon, iRobot's PackBot, and KumoTek's MK-705 Roosterbot.

ROBOTICS DESIGN INC

Robotics Design Inc. is a company that designs and builds modular robots, founded and incorporated in Montreal, Quebec, Canada in 1997. The company produces mobile robots, robotic manipulators and manual arms as well as custom solutions using modular robotic technology. The company developed the Bixi bike dock, a public bike system, and the ADC (Automatic Deployable Container), a deployable structure container for quick deployment of housing, hospitals and other buildings.

Robotics Design was founded in 1997 by Charles Khairallah M.ing after his invention of ANAT Technology. In 2000, Robotics Design Inc. released the first AMI-100 which was sold to the ETS university for research purposes. In 2003 Robotics Design Inc. became a member of the Canadian Intelligent Systems Companies. By 2004 the ANATROLLER line of mobile robots was introduced to the market with the arrival of the ARI-100 and, by 2006, the ARE-100 and ARI-50 joined the roster. Several new patents were introduced by Khairallah in 2008 with the release of BIXI which was named 19th best invention of 2008 by Time magazine, and winner of the 2009 Gold Edison award for Energy and sustainability.

Technology

Invented by Charles Khairallah in 1995, ANAT technology allows the creation of self-reconfiguring modular robots using U and H shaped modules, which are claimed to be more durable then previous L shaped modules. This allows any robot to be formed from connected identical modules, reducing the cost of manufacturing these products and increasing their work efficiency. It has led to the creation of several families of modular robots which have been used primarily for industrial purposes. ANAT technology was recognized as a nominee for the 2010 Manning Innovation Awards.

Mobile Robots

The ANATROLLER series of robots are a family of mobile robots based on modular ANAT technology. The first was the ANATROLLER ARI-100 which was launched in 2003 in North America. In 2010 the Kel'air duct cleaning company, a French company based in Bordeaux, acquired one of these, the first in the European market.

- The ANATROLLER ARI-100 is an industrial robot used primarily for duct cleaning. Weighing 8 kilograms (18 lb), it can climb obstacles of 13 centimetres (5.1 in), slopes of 45°, tow 45 kilograms (99 lb) and carry 50 kilograms (110 lb). It can support a number of accessories allowing it to be used for welding, nuclear reactor maintenance, and materials handling. These accessories can be attached to its articulated arm which can bend in four directions. Its symmetrical design allows it to continue driving when flipped on either side.
- The ANATROLLER ARI-50 is a smaller sized industrial duct cleaning robot, the ARI-50 can fit into 6 square inches (39 cm^2) spaces, climb obstacles and slopes, and carry loads. This robot's arm is stationary, although accessories can still be attached. Its modular design is made from half of the ARI-100, and can be connected to other ARI-50 robots like Lego blocks to form new robots.
- The ANATROLLER ARI-10 is a miniature duct cleaning robot specialized for residential duct cleaning with an advanced movement system that allows it to climb metallic surfaces vertically. It is equipped with HD cameras and a rugged metallic cable, and is fully customizable for special applications.
- The ANATROLLER ARE-100 is a wireless industrial robot whose base is constructed from a mono-block module and weighs 8 kilograms (18 lb). It can climb obstacles, slopes of 45° and tow 45 kilograms (99 lb). A basic modular mobile robot with an open architecture, the ARE-100 is used primarily for research purposes. Researchers can use the ARE-100 to develop specialized intelligent robotic projects for security applications. Such applications as human-machine interaction, mobile system navigation, robot behavior, image processing, object recognition, voice recognition, tele-operation, remote sensing, and autonomous navigation/patrol map building and localization.

Robot Manipulators

The AMI-100 is an industrial manipulator consisting of prismatic joints mounted on a fixed base, and a series of identical modules linked in an articulated SCARA configuration, which forms a snake-arm. It can be used for assembly, objects handling, maintenance, repair and further applications where a stationary arm with attached tools are used.

A portable tele-operated version of the AMI-100, the AEI-100 is an industrial manipulator, used primarily for materials handling from a distance. It is based on a modular redundant design which allows basic models to carry high payloads.

Manual Arms

The Anatergoarm AEA-15, winner of the gold medal at the 31st international Geneva Exhibition and a finalist in IDM, is a manual arm Deisgned in collaboration with Michel Dallaire Design Industrielle. It mimics the bio-mechanical movement of the human body which allows workers performing repetitive tasks to simplify their work, especially when using heavy tools. It incorporates an automatic safety brake system which allows the arm to remain stationary at will.

A portable Anatergoarm model, the TMA-500 is an assisted manipulator for maintenance and handling in of hydro-electric turbines. The first TMA-500 arm was deployed in 2010 at Hydro Quebec's Robert-Bourassa generating station for breaking unit repairs. It won the IRSST's 2011 "Work health and safety innovation" award at the 21st edition of the innovation awards organized by ADRIQ and its partners.

Special Projects

Bixi

Named 19th best invention of the year 2008 by *Time magazine*, Bixi utilizes Robotics Design Inc.'s modular docking station. Bixi is solar powered and can be removed and replaced quickly allowing entire stations to be deployed or removed in minutes. Bixi currently functions as the Montreal public bike system, and has also been sold internationally, from New York city to Melbourne.

ADC

ADC (Automated Deployable Container) is an invention that allows the transportation and deployment of habitable units. ADC helps humanitarian aid workers easily install housing, hospitals, and schools, making it possible to construct a temporary city in less than a week. The container is deployed automatically with a button control, and can change shape from a standard container compartment to a three compartment unit including kitchens and sanitary rooms. It is equipped with heating and air conditioning systems.

Guarded Tele-Op

A guarded tele-op robot has the ability to sense and avoid obstacles but will otherwise navigate as driven, like a robot under manual tele-op. Few if any mobile robots offer only guarded tele-op. *(See Sliding Autonomy below.)*

Line-Following Car

Some of the earliest Automated Guided Vehicles (AGVs) were line following mobile robots. They might follow a visual line painted or embedded in the floor or ceiling or an electrical wire in the floor. Most of these robots operated a simple "keep the line in the center sensor" algorithm. They could not circumnavigate obstacles; they just stopped and waited when something blocked their path. Many examples of such vehicles are still sold, by Transbotics, FMC, Egemin, HK Systems and many other companies. These types of robots are still widely popular in well known Robotic societies as a first step towards learning nooks and corners of robotics.

Autonomously Randomized Robot

Autonomous robots with random motion basically bounce off walls, whether those walls are sensed

Autonomously Guided Robot

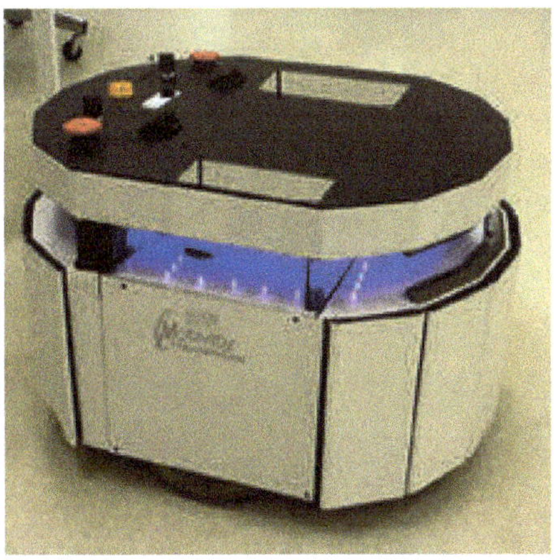

Robot developers use ready-made autonomous bases and software to design robot applications quickly. Shells shaped like people or cartoon characters may cover the base to disguise it. *Courtesy of MobileRobots Inc*

An autonomously guided robot knows at least some information about where it is and how to reach various goals and or waypoints along the way. "Localization" or knowledge of its current location, is calculated by one or more means, using sensors such motor encoders, vision, Stereopsis, lasers and global positioning systems. Positioning systems often use triangulation, relative position and/or Monte-Carlo/Markov localization to determine the location and orientation of the platform, from which it can plan a path to its next waypoint or goal. It can gather sensor readings that are time- and location-stamped. Such robots are often part of the wireless enterprise network, interfaced with other sensing and control systems in the building. For instance, the PatrolBot security robot responds to alarms, operates elevators and notifies the command center when an incident arises. Other autonomously guided robots include the SpeciMinder and the TUG delivery robots for the hospital. In 2013, autonomous robots capable of finding sunlight and water for potted plants were created by artist Elizabeth Demaray in collaboration with engineer Dr. Qingze Zou, biologist Dr. Simeon Kotchomi, and computer scientist Dr. Ahmed Elgammal.

Sliding Autonomy

More capable robots combine multiple levels of navigation under a system called sliding autonomy. Most autonomously guided robots, such as the HelpMate hospital robot, also offer a manual mode. The Motivity autonomous robot operating system, which is used in the ADAM, PatrolBot, SpeciMinder, MapperBot and a number of other robots, offers full sliding autonomy, from manual to guarded to autonomous modes.

Date	Developments
1939–1945	During World War II the first mobile robots emerged as a result of technical advances on a number of relatively new research fields like computer science and cybernetics. They were mostly flying bombs. Examples are smart bombs that only detonate within a certain range of the target, the use of guiding systems and radar control. Thc V1 and V2 rockets had a crude 'autopilot' and automatic detonation systems. They were the predecessors of modern cruise missiles.
1948–1949	W. Grey Walter builds Elmer and Elsie, two autonomous robots called Machina Speculatrix because these robots liked to explore their environment. Elmer and Elsie were each equipped with a light sensor. If they found a light source they would move towards it, avoiding or moving obstacles on their way. These robots demonstrated that complex behaviour could arise from a simple design. Elmer and Elsie only had the equivalent of two nerve cells.
1961–1963	The Johns Hopkins University develops 'Beast'. Beast used a sonar to move around. When its batteries ran low it would find a power socket and plug itself in.
1969	Mowbot was the very first robot that would automatically mow the lawn.
1970	The Stanford Cart line follower was a mobile robot that was able to follow a white line, using a camera to see. It was radio linked to a large mainframe that made the calculations. At about the same time (1966–1972) the Stanford Research Institute is building and doing research on Shakey the Robot, a robot named after its jerky motion. Shakey had a camera, a rangefinder, bump sensors and a radio link. Shakey was the first robot that could reason about its actions. This means that Shakey could be given very gencral commands, and that the robot would figure out the necessary steps to accomplish the given task. The Soviet Union explores the surface of the Moon with Lunokhod 1, a lunar rover.

1976	In its Viking program the NASA sends two unmanned spacecraft to Mars.
1980	The interest of the public in robots rises, resulting in robots that could be purchased for home use. These robots served entertainment or educational purposes. Examples include the RB5X, which still exists today and the HERO series. The Stanford Cart is now able to navigate its way through obstacle courses and make maps of its environment.
Early 1980s	The team of Ernst Dickmanns at Bundeswehr University Munich builds the first robot cars, driving up to 55 mph on empty streets.
1987	Hughes Research Laboratories demonstrates the first cross-country map and sensor-based autonomous operation of a robotic vehicle.
1989	Mark Tilden invents BEAM robotics.
1990s	Joseph Engelberger, father of the industrial robotic arm, works with colleagues to design the first commercially available autonomous mobile hospital robots, sold by Helpmate. The US Department of Defense funds the MDARS-I project, based on the Cybermotion indoor security robot.
1991	Edo. Franzi, André Guignard and Francesco Mondada developed Khepera, an autonomous small mobile robot intended for research activities. The project was supported by the LAMI-EPFL lab.
1993–1994	Dante I and Dante II were developed by Carnegie Mellon University. Both were walking robots used to explore live volcanoes.
1994	With guests on board, the twin robot vehicles VaMP and VITA-2 of Daimler-Benz and Ernst Dickmanns of UniBwM drive more than one thousand kilometers on a Paris three-lane highway in standard heavy traffic at speeds up to 130 km/h. They demonstrate autonomous driving in free lanes, convoy driving, and lane changes left and right with autonomous passing of other cars.
1995	Semi-autonomous ALVINN steered a car coast-to-coast under computer control for all but about 50 of the 2850 miles. Throttle and brakes, however, were controlled by a human driver.
1995	In the same year, one of Ernst Dickmanns' robot cars (with robot-controlled throttle and brakes) drove more than 1000 miles from Munich to Copenhagen and back, in traffic, at up to 120 mph, occasionally executing maneuvers to pass other cars (only in a few critical situations a safety driver took over). Active vision was used to deal with rapidly changing street scenes.
1995	The Pioneer programmable mobile robot becomes commercially available at an affordable price, enabling a widespread increase in robotics research and university study over the next decade as mobile robotics becomes a standard part of the university curriculum.
1996–1997	NASA sends the Mars Pathfinder with its rover Sojourner to Mars. The rover explores the surface, commanded from earth. Sojourner was equipped with a hazard avoidance system. This enabled Sojourner to autonomously find its way through unknown martian terrain.
1999	Sony introduces Aibo, a robotic dog capable of seeing, walking and interacting with its environment. The PackBot remote-controlled military mobile robot is introduced.
2001	Start of the Swarm-bots project. Swarm bots resemble insect colonies. Typically they consist of a large number of individual simple robots, that can interact with each other and together perform complex tasks.

2002	Roomba appears, a domestic autonomous mobile robot that cleans the floor.
2003	Axxon Robotics purchases Intellibot, manufacturer of a line of commercial robots that scrub, vacuum, and sweep floors in hospitals, office buildings and other commercial buildings. Floor care robots from Intellibot Robotics LLC operate completely autonomously, mapping their environment and using an array of sensors for navigation and obstacle avoidance.
2004	Robosapien, a biomorphic toy robot designed by Mark Tilden is commercially available. In 'The Centibots Project' 100 autonomous robots work together to make a map of an unknown environment and search for objects within the environment. In the first DARPA Grand Challenge competition, fully autonomous vehicles compete against each other on a desert course.
2005	Boston Dynamics creates a quadruped robot intended to carry heavy loads across terrain too rough for vehicles.
2006	Sony stops making Aibo and HelpMate halts production, but a lower-cost PatrolBot customizable autonomous service robot system becomes available as mobile robots continue the struggle to become commercially viable. The US Department of Defense drops the MDARS-I project, but funds MDARS-E, an autonomous field robot. TALON-Sword, the first commercially available robot with grenade launcher and other integrated weapons options, is released. Honda's Asimo learns to run and climb stairs.
2007	In the DARPA Urban Grand Challenge, six vehicles autonomously comple a complex course involving manned vehicles and obstacles. Kiva Systems robots proliferate in distribution operations; these automated shelving units sort themselves according to the popularity of their contents. The Tug becomes a popular means for hospitals to move large cabinets of stock from place to place, while the Speci-Minder with Motivity begins carrying blood and other patient samples from nurses' stations to various labs. Seekur, the first widely available, non-military outdoor service robot, pulls a 3-ton vehicle across a parking lot, drives autonomously indoors and begins learning how to navigate itself outside. Meanwhile, PatrolBot learns to follow people and detect doors that are ajar.
2008	Boston Dynamics released video footage of a new generation BigDog able to walk on icy terrain and recover its balance when kicked from the side.
2010	The Multi Autonomous Ground-robotic International Challenge has teams of autonomous vehicles map a large dynamic urban environment, identify and track humans and avoid hostile objects.
2016	The Multi-Function Agile Remote-Controlled Robot (MARCbot) is for the first time used by US police to kill a sniper who killed 5 police officers in Dallas, Texas, which raises ethical questions regarding the use of drones and robots by police as instruments of lethal force against a perpetrator. During the NASA Sample Return Robot Centennial Challenge, a rover, named Cataglyphis, successfully demonstrated autonomous navigation, decision-making, and sample detection, retrieval, and return capabilities.

AUTONOMOUS ROBOT

An autonomous robot is a robot that performs behaviors or tasks with a high degree of autonomy, which is particularly desirable in fields such as spaceflight, household maintenance (such as cleaning), waste water treatment and delivering goods and services.

Some modern factory robots are autonomous within the strict confines of their direct environment. It may not be that every degree of freedom exists in their surrounding environment, but the factory robot's workplace is challenging and can often contain chaotic, unpredicted variables. The exact orientation and position of the next object of work and (in the more advanced factories) even the type of object and the required task must be determined. This can vary unpredictably (at least from the robot's point of view).

One important area of robotics research is to enable the robot to cope with its environment whether this be on land, underwater, in the air, underground, or in space.

A fully autonomous robot can:

- Gain information about the environment
- Work for an extended period without human intervention
- Move either all or part of itself throughout its operating environment without human assistance
- Avoid situations that are harmful to people, property, or itself unless those are part of its design specifications

An autonomous robot may also learn or gain new knowledge like adjusting for new methods of accomplishing its tasks or adapting to changing surroundings.

Like other machines, autonomous robots still require regular maintenance.

Examples

Self-Maintenance

The first requirement for complete physical autonomy is the ability for a robot to take care of itself. Many of the battery-powered robots on the market today can find and connect to a charging station, and some toys like Sony's *Aibo* are capable of self-docking to charge their batteries.

Self-maintenance is based on proprioception, or sensing one's own internal status. In the battery charging example, the robot can tell proprioceptively that its batteries are low and it then seeks the charger. Another common proprioceptive sensor is for heat monitoring. Increased proprioception will be required for robots to work autonomously near people and in harsh environments. Common proprioceptive sensors include thermal, optical, and haptic sensing, as well as the Hall effect (electric).

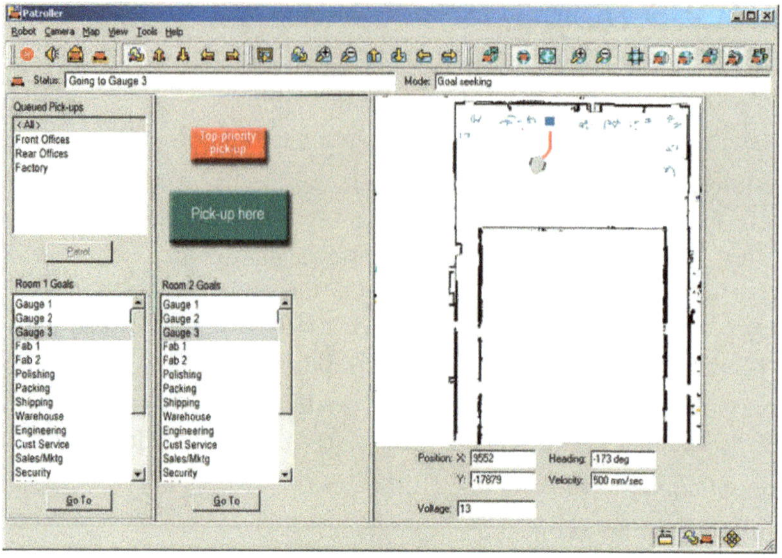

Robot GUI display showing battery voltage and other proprioceptive data in lower right-hand corner. The display is for user information only. Autonomous robots monitor and respond to proprioceptive sensors without human intervention to keep themselves safe and operating properly.

Sensing the Environment

Exteroception is sensing things about the environment. Autonomous robots must have a range of environmental sensors to perform their task and stay out of trouble.

- Common exteroceptive sensors include the electromagnetic spectrum, sound, touch, chemical (smell, odor), temperature, range to various objects, and altitude.

Some robotic lawn mowers will adapt their programming by detecting the speed in which grass grows as needed to maintain a perfectly cut lawn, and some vacuum cleaning robots have dirt detectors that sense how much dirt is being picked up and use this information to tell them to stay in one area longer.

Task Performance

The next step in autonomous behavior is to actually perform a physical task. A new area showing commercial promise is domestic robots, with a flood of small vacuuming robots beginning with iRobot and Electrolux in 2002. While the level of intelligence is not high in these systems, they navigate over wide areas and pilot in tight situations around homes using contact and non-contact sensors. Both of these robots use proprietary algorithms to increase coverage over simple random bounce.

The next level of autonomous task performance requires a robot to perform conditional tasks. For instance, security robots can be programmed to detect intruders and respond in a particular way depending upon where the intruder is.

Autonomous Navigation

Indoor Navigation

For a robot to associate behaviors with a place (localization) requires it to know where it is and to be able to navigate point-to-point. Such navigation began with wire-guidance in the 1970s and progressed in the early 2000s to beacon-based triangulation. Current commercial robots autonomously navigate based on sensing natural features. The first commercial robots to achieve this were Pyxus' HelpMate hospital robot and the CyberMotion guard robot, both designed by robotics pioneers in the 1980s. These robots originally used manually created CAD floor plans, sonar sensing and wall-following variations to navigate buildings. The next generation, such as MobileRobots' PatrolBot and autonomous wheelchair, both introduced in 2004, have the ability to create their own laser-based maps of a building and to navigate open areas as well as corridors. Their control system changes its path on the fly if something blocks the way.

At first, autonomous navigation was based on planar sensors, such as laser range-finders, that can only sense at one level. The most advanced systems now fuse information from various sensors for both localization (position) and navigation. Systems such as Motivity can rely on different sensors in different areas, depending upon which provides the most reliable data at the time, and can re-map a building autonomously.

Rather than climb stairs, which requires highly specialized hardware, most indoor robots navigate handicapped-accessible areas, controlling elevators, and electronic doors. With such electronic access-control interfaces, robots can now freely navigate indoors. Autonomously climbing stairs and opening doors manually are topics of research at the current time.

As these indoor techniques continue to develop, vacuuming robots will gain the ability to clean a specific user-specified room or a whole floor. Security robots will be able to cooperatively surround intruders and cut off exits. These advances also bring concomitant protections: robots' internal maps typically permit forbidden areas to be defined to prevent robots from autonomously entering certain regions.

Outdoor Navigation

Outdoor autonomy is most easily achieved in the air, since obstacles are rare. Cruise missiles are rather dangerous highly autonomous robots. Pilotless drone aircraft are increasingly used for reconnaissance. Some of these unmanned aerial vehicles (UAVs) are capable of flying their entire mission without any human interaction at all except possibly for the landing where a person intervenes using radio remote control. Some drones are capable of safe, automatic landings, however. An autonomous ship was announced in 2014—the Autonomous spaceport drone ship—and is scheduled to make its first operational test in December 2014.

Outdoor autonomy is the most difficult for ground vehicles, due to:

- Three-dimensional terrain
- Great disparities in surface density

- Weather exigencies
- Instability of the sensed environment

The Seekur and MDARS robots demonstrate their autonomous navigation and security capabilities at an airbase. (Courtesy of Omron Adept)

In the United States, the MDARS project, which defined and built a prototype outdoor surveillance robot in the 1990s, is now moving into production and will be implemented in 2006. The General Dynamics MDARS robot can navigate semi-autonomously and detect intruders, using the MRHA software architecture planned for all unmanned military vehicles. The Seekur robot was the first commercially available robot to demonstrate MDARS-like capabilities for general use by airports, utility plants, corrections facilities and Homeland Security.

The Mars rovers MER-A and MER-B (now known as Spirit rover and Opportunity rover) can find the position of the sun and navigate their own routes to destinations on the fly by:

- Mapping the surface with 3D vision
- Computing safe and unsafe areas on the surface within that field of vision
- Computing optimal paths across the safe area towards the desired destination
- Driving along the calculated route;
- Repeating this cycle until either the destination is reached, or there is no known path to the destination

The planned ESA Rover, ExoMars Rover, is capable of vision based relative localisation and absolute localisation to autonomously navigate safe and efficient trajectories to targets by:

- Reconstructing 3D models of the terrain surrounding the Rover using a pair of stereo cameras

- Determining safe and unsafe areas of the terrain and the general "difficulty" for the Rover to navigate the terrain
- Computing efficient paths across the safe area towards the desired destination
- Driving the Rover along the planned path
- Building up a navigation map of all previous navigation data

During the final NASA Sample Return Robot Centennial Challenge in 2016, a rover, named Cataglyphis, successfully demonstrated fully autonomous navigation, decision-making, and sample detection, retrieval, and return capabilities. The rover relied on a fusion of measurements from inertial sensors, wheel encoders, Lidar, and camera for navigation and mapping, instead of using GPS or magnetometers. During the 2 hour challenge, Cataglyphis traversed over 2.6 km and returned five different samples to its starting position.

The DARPA Grand Challenge and DARPA Urban Challenge have encouraged development of even more autonomous capabilities for ground vehicles, while this has been the demonstrated goal for aerial robots since 1990 as part of the AUVSI International Aerial Robotics Competition.

Open Problems in Autonomous Robotics

There are several open problems in autonomous robotics which are special to the field rather than being a part of the general pursuit of AI. According to George A. Bekey's *Autonomous Robots: From Biological Inspiration to Implementation and Control*, problems include things such as making sure the robot is able to function correctly and not run into obstacles autonomously.

Energy Autonomy and Foraging

Researchers concerned with creating true artificial life are concerned not only with intelligent control, but further with the capacity of the robot to find its own resources through foraging (looking for food, which includes both energy and spare parts).

This is related to autonomous foraging, a concern within the sciences of behavioral ecology, social anthropology, and human behavioral ecology; as well as robotics, artificial intelligence, and artificial life.

MOBILE INDUSTRIAL ROBOTS

Mobile industrial robots are pieces of machinery that are able to be programmed to perform tasks in an industrial setting. Typically these have been used in stationary and workbench applications; however, mobile industrial robots introduce a new method for lean manufacturing. With advances in controls and robotics, current technology has been improved allowing for mobile tasks such as product delivery. This additional flexibility in manufacturing can save a company time and money during the manufacturing process, and therefore results in a cheaper end product. Mobile robot technology has potential to revolutionize many sectors of industry; however, it carries with it some disadvantages. The logistics of manufacturing will be streamlined by allowing robots to autonomously navigate to different areas for their work. The labor demands for employees will be lessened as robots will be able to work alongside humans, and robots will assist with medicine and surgery more and more. However, there are drawbacks to this technology. Coordinating the movement of robots around facilities and calibrating their position at their destination is tedious and far from perfect. A robot malfunctioning in a manufacturing setting will hold up production - and this robot could malfunction anywhere in a facility. Human safety must also be considered. Robots must prioritize the safety of human operators over their programmed task - which may complicate the coordination of multiple autonomous robots. Especially in a surgical setting, there is no room for error on the robot's part. Even though some challenges are present, mobile robot technology promises to streamline aspects across much of the industry.

Automation began in the automobile industry in the years surrounding WWII (1946) and the origin of the term itself belongs with D.S. Harder, the engineering manager at the Ford Motor Company. At first, the term was used to describe the increased presence of automatic devices in production lines and solely manufacturing contexts. Now, automation is widely used in many industries where computerized action and feedback loops can replace human intervention in the workplace. Over time, development in this area has become increasingly dependent upon advanced computer technologies and the advancement of processing capabilities.

In its current form, most industrial robots are powered mechanical arms with the ability to perform anthropomorphic actions. Advancements in miniaturization of computers, mathematical control theory as well as improved sensory technologies have had great impact on the feedback control systems that drive robotics. The first industrial robot performed spot welding and die castings in a General Motors factory in New Jersey, USA in 1962. Soon, robotic arms were exploding within the large-scale manufacturing industry and several new companies came into existence including Kuka in 1973, Nachi in 1969, Fanuc in 1974, Yaskawa in 1977, ASEA in 1977, and several others. By 1980, it is estimated a new major robotics company entered the market every month.

Mobile robotics are now set to experience similar expansion as they become significantly more reliable in an industrial setting. Even if a mobile robot makes mistakes, it will eventually be less frequently than mistakes caused by human factors.

Overview

The simplicity of mobile industrial robots provide their main advantage in industrial settings due to the ease of use and ability to be operated via technologies well understood by most people. In addition, robots are able to operate almost continuously and will never complain about long work hours; greatly increasing efficiency in a lean manufacturing environment. The main current disadvantage lies in high costs of repair as well as the production delays that would be caused by a failure or malfunction. These factors are very preventative to putting major amounts of responsibility on mobile robotics, however they are being continually lessened.

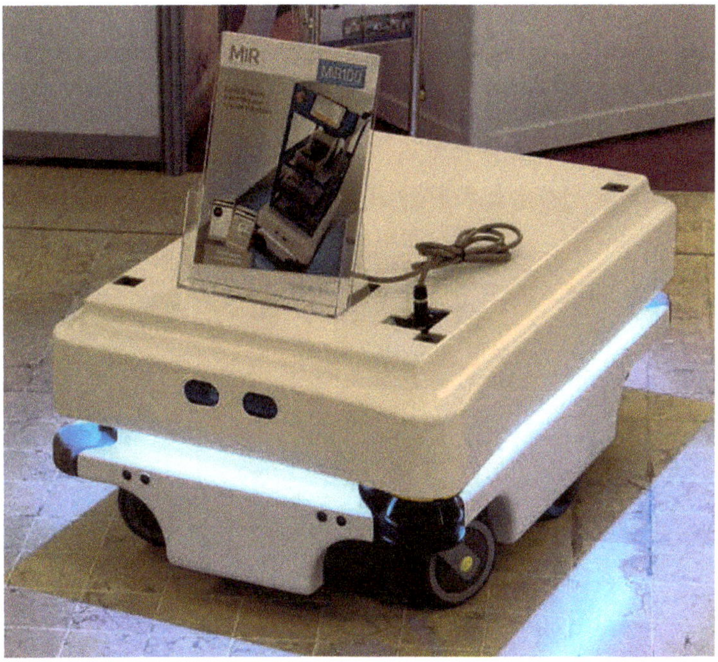

MiR100, a frontrunning example product for mobile industrial robots.

Applications of Mobile Industrial Robots

The mobile industrial robots have many applications that they have been used in already including in the healthcare industry, home and industrial security, ocean and space exploration, the food service industry, and in distribution applications.

Medicine

Mobile industrial robots have several uses within the healthcare industry in both hospitals and homes. Drug delivery, patient services, and other nursing functions could be easily adapted to robots. Due to the fact that items being carried around typically weigh less than 100 kg, robots much smaller than the MiR (see above) may be used. Specialized equipment may be mounted on robots, allowing them to assist with surgical procedures. Overall, their place in the medical industry is to provide a more reliable source of customer care while reducing human error.

Scientific Experimentation and Exploration

In the scientific world, there is a large number of applications for mobile robots. Their ability to perform experiments and exploration without putting human lives in danger makes them an important asset. Unlike humans, robots do not require life support systems to function. In space travel, robots are performing science on planets and asteroids because sending humans is far more taxing on resources and money. The same is true in the oceanography domain. In fact, several of the same robotic systems are designed to perform their science under both conditions - space and underwater. In nuclear power plants, robots can service electronics and mechanical systems which prevents human exposure to large amounts of radiation.

Aircraft Maintenance and Repair

Air-Cobot is a collaborative mobile robot able to inspect aircraft. Picture of the robot in Air France Industries.

For applications like painting and de-painting aircraft, two fixed robots are inadequate because not all parts of the aircraft can be reached. Adding more fixed robots would complete the task, but the cost is prohibitive. If mobile robots are used, one or two may be enough to service the entire aircraft because they can move to whatever area needs work. Mobile robots need to be truly autonomous to be useful in manufacturing. Erik Nieves said "Mobility moves robots from being machines to production partners " Rather than bringing work to the robot, the robot should be smart enough to go to where the work is.

Automated aircraft inspection systems have the potential to make aircraft maintenance safer and more reliable. Various solutions are currently developed: a collaborative mobile robot named Air-Cobot,and drones.

Examples

OTTO Motors (a Division of Clearpath)

OTTO 1500 self-driving vehicle for heavy-load material transport in warehouses, distribution centers, and factories.

Meant for material transport in industrial centers

- Able to carry 100 kg (OTTO 100) or 1500 kg (OTTO 1500)
- Powered by lithium battery technology
- 6-8 hours operating time (depending on payload)
- Requires no structural changes to building
- Navigates via 2D sensors. Proprietary autonomy software enables dynamic path-planning and obstacle avoidance at speeds matching forklifts (2 m/s).
- Used by GE and John Deere.

MiR (Mobile Industrial Robots Corp.)

- Meant for transportation
    - Able to carry 100 kg, large enough to hold full size standard EU pallets
    - Electrically powered
    - Long operating time on a single battery charge (12–15 hours/20 km)
    - Maximum velocity: 1.5 m/s

- Requires no structural changes of building (i.e. wires in the floor, sensors in the ceiling)
- Can be operated via smartphone or PC
- Navigates via an array of 2D and 3D scanners and cameras

KUKA youBot

Kuka

- Very widely used—Example: Tesla Motors
- "Mecanum" wheel system: customizeable, modular, heavy-lifting capable
- Very easy to integrate with autonomous robotics and humans

REFERENCES

1. Optically Automated Spy Robot, 'OASR', Gaurav Mittal and Deepansh Sehgal, Punjab Engineering College
2. P. Moubarak, P. Ben-Tzvi, Adaptive Manipulation of a Hybrid Mechanism Mobile Robot, IEEE International Symposium on Robotic and Sensors Environments (ROSE), Montreal, Canada, 2011, pp. 113 - 118
3. Gopalakrishnan, B.; Tirunellayi, S.; Todkar, R. (2014). "Design and development of an autonomous mobile smart vehicle: a mechatronics application". Mechatronics. 14: 491–514.
4. Rail track and Linear track (PDF) Archived July 15, 2011, at the Wayback Machine.
5. Archived February 23, 2010, at the Wayback Machine.
6. WOOLLASTON, VICTORIA (9 May 2013). "Plant bot: The world's first robot that can turn your household plants into light-seeking 'triffid' drones". Dailymail.

co.uk. Retrieved 9 January 2014.

7. "ias-people". Ias.uwe.ac.uk. Retrieved 2012-08-15.
8. Proceedings of IEEE Robotics and Automation, 1988
9. "Robotics Institute: Dante I". Ri.cmu.edu. Retrieved 2012-08-15.
10. "Robotics Institute: Dante II". Ri.cmu.edu. Retrieved 2012-08-15.
11. "Centibots Project Home Page". Ai.sri.com. 2004-10-04. Retrieved 2012-08-15.
12. Archived December 6, 2006, at the Wayback Machine.
13. Welcome [https://web.archive.org/web/20080416113845/http://www.darpa.mil/GRANDCHALLENGE/ Archived April 16, 2008, at the Wayback Machine.
14. Archived September 28, 2007, at the Wayback Machine.
15. http://heavy.com/news/2016/07/dallas-swat-robot-bomb-kill-micah-x-johnson-first-lethal-use-explosive/
16. Hall, Loura (2016-09-08). "NASA Awards $750K in Sample Return Robot Challenge". Retrieved 2016-09-21.
17. "The Tank Tread Was His Baby". Popular Science (June): 63. 1944. Retrieved 2011-08-24.
18. "Josef-Maria Hoëné de Wronski". Retrieved 2009-05-30.
19. "Sir George Cayley's patent universal railway". Mechanics Magazine. 5 (127): 225–227. 1826-01-28.
20. "Burrell's traction engine with Boydell's endless railway". Grace's Guide. 1857. Retrieved 2013-09-30.
21. Lane, Michael R. (1994). The story of the St. Nicholas Works: A History of Charles Burrell and Sons. London: Unicorn Press. ISBN 978-0906290071.
22. "Boydell Artillery Wheel" (PDF). Retrieved 2013-09-30.
23. "Governments Experiments with Boydell's Traction-Engine". The Farmer's Magazine. London. 45 (1). 1856-06-30.
24. "Tuxford's Boydell Traction Engine" (PDF). Science& Society: Picture Library. UK. 1857. Retrieved 2014-02-04.
25. op. cit. Michael R. Lane 1994
26. Clark, Ronald H. (1974). The Development of the English Traction Engine. Cambridge UK: Goose and Son. ISBN 0900404027.
27. "Charles Burrell & Sons Limited" (PDF). University of Reading. UK. Retrieved 2013-09-30.
28. "Burrell-Boydell Tractor" (PDF). Retrieved 2013-09-30.
29. Caley, Georg (1825). "Patent No. 5260 A New Locomotive Apparatus" (PDF). Retrieved 2013-09-30.
30. Lozovoi, D.; Lozova, A. "Изобретатель трактора (Ф. Блинов) [The inventor of the tractor (F.Blinov)]. Russia in Colours (in Russian). Retrieved 2011-08-24.
31. The Implement and Machinery Review, 1901-01-02

32. Invention of the Caterpillar Track: Frank Beamond and his Patents (PDF). Retrieved 2014-12-10.
33. Lombard Log Hauler and Model T Snowmobile Show Route 175 Thornton, NH. Retrieved 2011-08-24.
34. British Patent No. 16,345 (1904)
35. Simkin, John. Little Willie Tank. UK: Spartacus Educational. Retrieved 2016-11-08.
36. About Caterpillar. US: Antique Caterpillar Machinery Owners Club. Retrieved 2016-11-07.
37. RF Scott (1908) The Sledging Problem in the Antarctic, Men versus Motors
38. Roland Huntford (2003) Scott and Amundsen. Their Race to the South Pole. The Last Place on Earth. Abacus, London, p224
39. riverbanksy (2011-08-21). 90 Degrees South. Retrieved 2016-10-23 – via YouTube.
40. Cherry-Garrard, Apsley (1970) [1922]. The Worst Journey in the World chapter 9. London: Penguin Books. ISBN 978-0-14-009501-2. OCLC 16589938.
41. Haddock, Keith. Giant Earthmovers: An Illustrated History. MotorBooks International. pp. 17, 20, 21. Retrieved 2015-08-16.
42. The Earthmover Encyclopedia, Page 28, Keith Haddock
43. US patent 69987, James K Glen, Improvement in Motive Power, issued 1867-10-22
44. US patent 373887, William Fender, Wheel With Endless Rail, issued 1887-11-29
45. US patent 433488, Goldsbury Harden Pond, Traction Engine, issued 1890-08-05
46. Scientific American, December 18, 1886, Vol. LV, No. 25
47. Kane, Joseph Nathan, Famous First Facts, H. W. Wilson Company (1950), p. 47
48. US patent 351749, Charles Dinsmoor, Vehicle, issued 1886-11-02A design for a tracked vehicle.
49. Caterpillar Tractor Co. List of Deals. Lehman Brothers Collection. President and Fellows of Harvard College. 2010. Archived from the original on 2010-06-29. Retrieved 2010-11-06. In 1925 Holt and C.W. Best's company merged to form the Caterpillar Tractor Company.
50. Austenitic Manganese Steels. Retrieved 2011-08-24.
51. Peter Chamberlain and Hilary Doyle, Encyclopedia of German Tanks of World War II, 1999
52. Official website. Liebherr. Retrieved 2013-05-03.
53. Kubota crawler tractor. Kubota.com. 2008-07-14. Retrieved 2013-05-03.
54. Search results for Used Tracked tractors. Mascus.co.uk. Retrieved 2013-05-03.
55. Burner, Ken (1997). The Small Tractor FAQ - Tractor Resources. Carnegie Mellon University. Archived from the original on 2012-06-22.
56. The Zavolzhsky Crawler Vehicle Plant. Russia: Zavolzhsky Crawler Vehicle

Plant. Retrieved 2016-11-07.

57. Main. Russia: Vityaz machine-building company. Retrieved 2016-11-07.
58. Edison Awards 2010 honoring innovation in the development and launch of new products and services. Edisonawards.com. April 1, 2009. Retrieved December 7, 2016.
59. Centech promotional brochure (pdf) (Press release). centech.etsmtl.ca. Retrieved December 20, 2009.
60. Roberts, Susan (January 2013). Robotics Design Builds Nontraditional Robots One Module at a Time Imagination Drives Design. FMA Communications, Inc.,. Retrieved January 17, 2013.
61. Technology. Roboticsdesign.qc.ca. Retrieved September 28, 2011.
62. Scanlan, Steve, Robotics Design Inc., Montreal (November 2009). Modular Robotics. Rogers Publishing. Retrieved November 9, 2009.
63. Deschamps, Ing., Isabelle (August–September 1997). Un belle Example D'innovation (PDF). roboticsdesign.qc.ca/news.html (in French). Ordre des Ingénieurs du Québec. Retrieved October 18, 2009.
64. News & Events Robotics Design Inc
65. Sud Ouest, Magazine (December 6, 2010). Climatisations et locaux confinées : de l'air pour nos intérieurs. http://www.sudouest.fr/ (in French). http://www.sudouest.fr/. Retrieved February 4, 2011.External link in |publisher=, |work= (help)
66. The ARI-100. Roboticsdesign.qc.ca. Retrieved October 16, 2009.
67. Construire, Magazine (March–April 2007). L'ANATROLLER fait le ménage (PDF). roboticsdesign.qc.ca/news.html (in French). http://www.acq.org. Retrieved December 7, 2009. External link in |publisher= (help)
68. Plumbing and HVAC, Magazine (April 2010). Duct cleaning robots(PDF). roboticsdesign.qc.ca/news.html. http://plumbingandhvac.ca/. Retrieved April 29, 2010. External link in |publisher= (help)
69. The ARI-50. Roboticsdesign.qc.ca. Retrieved October 16, 2009.
70. ARI-10. Roboticsdesign.ca. Retrieved September 7, 2012.
71. The ARE-100. Roboticsdesign.qc.ca. Retrieved October 16, 2009.
72. AMI-100. Roboticsdesign.qc.ca. Retrieved October 16, 2009.
73. Scanlan, Steve, Robotics Design Inc., Montreal. Modularity in robotics provides automation for all (PDF). Digital.ept.ca. Retrieved December 7, 2016.
74. Technology. Roboticsdesign.qc.ca. Retrieved October 16, 2009.
75. Des designers québécois remportent des Prix à Genève (pdf)(Press release). Retrieved December 20, 2009.
76. Prix de l'IDM 2003 Finalistes. Infodesigncanada.com. Retrieved October 16, 2009.
77. Montreal, Robotics Design Inc. (November 2009). Flexible Robot Eases Mould Changes (PDF). Rogers Publishing. Retrieved December 23, 2009.
78. AEA-15. Roboticsdesign.qc.ca. Retrieved October 16, 2009.

79. Montreal, Robotics Design Inc. (May 26, 2011). Arm Allows Brake Unit Repair at Hydro-Quebec Facility. PennWell Publishing. Retrieved Dec 7, 2016.
80. The Prix Innovation SST Awarded to Robotics Design (PDF). http://www.adriq.com/. Retrieved September 7, 2012. External link in |publisher= (help)
81. 19. Montreal's Public Bike System – TIME's Best Inventions of 2008. TIME. October 29, 2008. Retrieved October 16, 2009.
82. Marc Tison. Le Volksvélo (PDF). La Presse. Retrieved September 28, 2008.
83. Robotics Design Inc. News & Events. Accessed October 2011.
84. Lamoureux, Sébastien (November 2009). Robotics Design à l'origine d'innovations audacieuses du Bixi (in French). ÉTS. Retrieved November 6, 2009.
85. BIXI, la création québécoise sur deux roues (PDF). roboticsdesign.qc.ca (in French). http://montrealmetropoleculturelle.org/.
86. How America's biggest bike share will turn NYC into a cycling city.. GIZMODO. May 29, 2013. Retrieved June 12, 2013.
87. Principal Investigator: W. Kennedy, National Institutes of Health, NIH SBIR 2 R44 HD041781-02
88. Speci-Minder; see elevator and door access Archived January 2, 2008, at the Wayback Machine.
89. Bergin, Chris (2014-11-18). Pad 39A – SpaceX laying the groundwork for Falcon Heavy debut. NASA Spaceflight. Retrieved 2014-11-17.
90. Weapons Makers Unveil New Era of Counter-Terror Equipment, Fox News
91. Hall, Loura (2016-09-08). NASA Awards $750K in Sample Return Robot Challenge. Retrieved 2016-09-17
92. Zhang, Biao; Martinez, C.; Wang, Jianjun; Fuhlbrigge, T.; Eakins, W.; Chen, Heping (2010-08-01). The challenges of integrating an industrial robot on a mobile platform. 2010 IEEE International Conference on Automation and Logistics (ICAL): 255–260. doi:10.1109/ICAL.2010.5585289.
93. Why Should we use Autonomous Industrial Mobile Manipulators - Smashing Robotics. Smashing Robotics. Retrieved 2016-03-16.
94. automation. Encyclopedia Britannica. Retrieved 2016-03-16.
95. Industrial Robot History. www.robots.com. Retrieved 2016-03-16.
96. Robotics Online. Robotics Online. Retrieved 2016-03-16.
97. Robots in the hangar. November 23, 2015. Retrieved May 20, 2016.
98. Air-Cobot. Retrieved May 20, 2016.
99. Jovancevic, Igor; Larnier, Stanislas; Orteu, Jean-José; Sentenac, Thierry (November 2015). Automated exterior inspection of an aircraft with a pan-tilt-zoom camera mounted on a mobile robot. Journal of Electronic Imaging. 24 (6).
100. I. Jovancevic, I. Viana, T. Sentenac, J.J. Orteu and S. Larnier, Matching CAD model and images features for robot navigation and inspection of an aircraft, International Conference on Pattern Recognition Applications and Methods, pp.

359-366, February 2016.

101. Donecle - lightning fast aircraft inspections. Retrieved May 20, 2016.
102. EasyJet's Using Drones to Check Planes for Lightning Damage. Retrieved May 20, 2016.
103. OTTO Motors. Retrieved Aug 8, 2016.
104. Clearpath to Provide GE Healthcare Repair Center with Self-Driving Vehicles. Retrieved Aug 9, 2016.
105. Clearpath Joins John Deere Supply Base. February 23, 2016.
106. Mobile Industrial Robots | Robot til transportopgaver i industrien og i sundhedssektoren.. mobile-industrial-robots.com. Retrieved 2016-03-16.
107. Mobile Industrial Robots ApS (2014-12-10), MiR at Hannover 2014, retrieved 2016-03-16
108. KUKA Industrial Robots - Mobility. www.kuka-robotics.com. Retrieved 2016-03-16.

CHAPTER 7

Beam Robotics

Chapter Contents

Common robots use a programmable microcontroller, which means some kind of computer, to make decisions. The more complex the behavior of the robot the more complex the program in the microcontroller. Anyone who has already had some experience programming a robot can acknowledge this. With every new task the robot has to perform the complexity of the robot rises exponentially.

INTRODUCTION

BEAM robotics (from *Biology, Electronics, Aesthetics and Mechanics*) is a style of robotics that primarily uses simple analogue circuits, such as comparators, instead of a microprocessor in order to produce an unusually simple design. While not as flexible as microprocessor based robotics, BEAM robotics can be robust and efficient in performing the task for which it was designed.

BEAM robots may use a set of the analog circuits, mimicking biological neurons, to facilitate the robot's response to its working environment.

MECHANISMS AND PRINCIPLES

The basic BEAM principles focus on a stimulus-response based ability within a machine. The underlying mechanism was invented by Mark W. Tilden where the circuit (or a Nv net of Nv neurons) is used to simulate biological neuron behaviours. Some similar research was previously done by Ed Rietman in 'Experiments In Artificial Neural Networks›. Tilden›s circuit is often compared to a shift register, but with several important features making it a useful circuit in a mobile robot.

Other rules that are included (and to varying degrees applied):

1. Use the lowest number possible of electronic elements (*"keep it simple"*)
2. Recycle and reuse technoscrap
3. Use radiant energy (such as solar power)

There are a large number of BEAM robots designed to use solar power from small solar arrays to power a "Solar Engine" which creates autonomous robots capable of operating under a wide range of lighting conditions. Besides the simple computational layer of Tilden's "Nervous Networks", BEAM has brought a multitude of useful tools to the roboticist's toolbox. The "Solar Engine" circuit, many H-bridge circuits for small motor control, tactile sensor designs, and meso-scale (palm-sized) robot construction techniques have been documented and shared by the BEAM community.

BEAM Robots

Being focused on "reaction-based" behaviors (as originally inspired by the work of Rodney Brooks), BEAM robotics attempts to copy the characteristics and behaviours of biological organisms, with the ultimate goal of domesticating these "wild" robots. The aesthetics of BEAM robots derive from the principle "form follows function" modulated by the particular design choices the builder makes while implementing the desired functionality.

Disputes in the Name

Various people have varying ideas about what BEAM actually stands for. The most widely accepted meaning is ***B**iology, **E**lectronics, **A**esthetics, and **M**echanics.*

This term originated with Mark Tilden during a discussion at the Ontario Science Centre in 1990. Mark was displaying a selection of his original bots which he had built while working at the University of Waterloo.

However, there are many other semi-popular names in use, including:

- **B**iotechnology **E**thology **A**nalogy **M**orphology
- **B**uilding **E**volution **A**narchy **M**odularity

Microcontrollers

Unlike many other types of robots controlled by microcontrollers, BEAM robots are built on the principle of using multiple simple behaviours linked directly to sensor systems with little signal conditioning. This design philosophy is closely echoed in the classic book "Vehicles: Experiments in Synthetic Psychology". Through a series of thought experiments, this book explores the development of complex robot behaviours through simple inhibitory and excitory sensor links to the actuators. Microcontrollers and computer programming are usually not a part of a traditional (aka., "pure") BEAM robot due to the very low-level hardware-centric design philosophy.

There are successful robot designs mating the two technologies. These "hybrids" fulfill a need for robust control systems with the added flexibility of dynamic programming, like the "horse-and-rider" topology BEAMbots (e.g. the ScoutWalker 3). 'Horse' behavior is implemented with traditional BEAM technology but a microcontroller based 'rider' can guide that behavior so as to accomplish the goals of the 'rider'.

Types

There are various "*-trope*" BEAMbots, which attempt to achieve a specific goal. Of the series, the phototropes are the most prevalent, as light-seeking would be the most beneficial behaviour for a solar-powered robot.

- Audiotropes react to sound sources.
  - *Audiophiles* go towards sound sources.
  - *Audiophobes* go away from sound sources.
- Phototropes ("light-seekers") react to light sources.
  - *Photophiles* (also *Photovores*) go toward light sources.
  - *Photophobes* go away from light sources.
- Radiotropes react to radio frequency sources.
  - *Radiophiles* go toward RF sources.
  - *Radiophobes* go away from RF sources.
- Thermotropes react to heat sources.
  - *Thermophiles* go toward heat sources.
  - *Thermophobes* go away from heat sources.

Audiotrope

In BEAM robotics, an audiotrope is a robot that reacts to sounds. This term, which literally means "sound turning," is generally applied to sound-seeking robots. More accurately, audiotropes can either seek (audiophiles) or avoid (audiophobes) sources of sound.

Phototrope

In BEAM robotics, a **Phototrope** is a robot that reacts to light sources. Literally, "light turning," this term is generally (if somewhat inaccurately) applied to light-seeking robots. More accurately, phototropes can either seek (photophiles) or flee (photophobes) bright sources of light.

The simplest and most common form of phototrope is the photopopper—many are as simple as to be essentially two solarrollers stuck together.

One mechanism for phototropism in robotics is implementation of a light sensor where a direct feedback system allows for interaction with the environment. The phototrope analyzes «shots» of its environment and decides whether to move into a certain area depending upon the light intensity.

Alternatively, photovoltaic cells may be used to provide both control and energy for a phototrope. Clever geometry in construction allows for current yielded by a photovoltaic cell to cause motion in the direction of (or away from) the most intense light source in the robots vicinity.

Radiotrope

In BEAM robotics, a **radiotrope** is a robot that reacts to radio frequency sources. Literally, "radio turning," this term is generally (if somewhat inaccurately) applied to radio frequency seeking robots. More accurately, radiotropes can either seek (radiophiles) or flee (radiophobes) sources of radio waves.

Thermotrope

In BEAM robotics, a **thermotrope** is a robot that reacts to heat sources. Literally, "heat turning," this term is generally (if somewhat inaccurately) applied to heat-seeking robots. More accurately, thermotropes can either seek (thermophiles) or flee (thermophobes) sources of heat.

General

BEAMbots have a variety of movements and positioning mechanisms. These include:

- *Sitters*: Unmoving robots that have a physically passive purpose.
  - Beacons: Transmit a signal (usually a navigational blip) for other BEAMbots to use.
  - Pummers: Display a "light show".
  - Ornaments: A catch-all name for sitters that are not beacons or pummers.
- *Squirmers*: Stationary robots that perform an interesting action (usually by moving some sort of limbs or appendages).
  - Magbots: Utilize magnetic fields for their mode of animation.

- Flagwavers: Move a display (or “flag”) around at a certain frequency.
- Heads: Pivot and follow some detectable phenomena, such as a light (These are popular in the BEAM community. They can be stand-alone robots, but are more often incorporated into a larger robot.).
- Vibrators: Use a small pager motor with an off-centre weight to shake themselves about.

■ *Sliders*: Robots that move by sliding body parts smoothly along a surface while remaining in contact with it.

- Snakes: Move using a horizontal wave motion.
- Earthworms: Move using a longitudinal wave motion.

■ *Crawlers*: Robots that move using tracks or by rolling the robot’s body with some sort of appendage. The body of the robot is not dragged on the ground.

- Turbots: Roll their entire bodies using their arm(s) or flagella.
- Inchworms: Move part of their bodies ahead, while the rest of the chassis is on the ground.
- Tracked robots: Use tracked wheels, like a tank.

■ *Jumpers*: Robots which propel themselves off the ground as a means of locomotion.

- Vibrobots: Produce an irregular shaking motion moving themselves around a surface.
- Springbots: Move forward by bouncing in one particular direction.

■ *Rollers*: Robots that move by rolling all or part of their body.

- Symets: Driven using a single motor with its shaft touching the ground, and moves in different directions depending on which of several symmetric contact points around the shaft are touching the ground.
- Solarrollers: Solar-powered cars that use a single motor driving one or more wheels; often designed to complete a fairly short, straight and level course in the shortest amount of time.
- Poppers: Use two motors with separate solar engines; rely on differential sensors to achieve a goal.
- Miniballs: Shift their centre of mass, causing their spherical bodies to roll.

■ *Walkers*: Robots that move using legs with differential ground contact.

- Motor Driven: Use motors to move their legs (typically 3 motors or less).
- Muscle Wire Driven: Utilise Nitinol (nickel - titanium alloy) wires for their

leg actuators.

- *Swimmers*: Robots that move on or below the surface of a liquid (typically water).
  - Boatbots: Operate on the surface of a liquid.
  - Subbots: Operate under the surface of a liquid.
- *Fliers*: Robots that move through the air for sustained periods.
  - Helicopters: Use a powered rotor to provide both lift and propulsion.
  - Planes: Use fixed or flapping wings to generate lift.
  - Blimps: Use a neutrally-buoyant balloon for lift.
- *Climbers*: Robot that moves up or down a vertical surface, usually on a track such as a rope or wire.

SITTER (BEAM)

In BEAM robotics, a **sitter** is a type of robot that does not move. Such robots have no self-contained means of transportation and are non-mobile. Instead of moving, sitters are usually designed to power lights in some sort of pattern. Most sitters are built as decorations rather than serving a practical purpose.

As sitters usually have few parts and are not very complicated to build, many newcomers to BEAM robotics start with sitters, as they are a good place to begin learning the basics of the hobby.

Genera

- Beacons : Transmit a signal (usually a navigational blip) for other BEAMbots to use, often as a navigational device. Essentially, these consist of a solar cell, capacitor, LED(s), and possibly a solar engine. Beacons are often found within robot games and are used by other robots to use as a "base" for navigation.
- Pummers : Display a «light show» or a pattern of sounds. Pummers are often nocturnal robots that store solar energy during the day, then activate during the night.
- Ornaments : A catch-all name for sitters which are not beacons or pummers. Many times, these are mostly electronic art.

SQUIRMER (BEAM)

In BEAM robotics, a **squirmer** is a stationary robot that performs an interesting action (usually by moving some sort of limbs or appendages).

Genera

- Magbots: Utilize magnetic fields for their mode of animation.
- Flagwavers: Move a display (or "flag") around at a certain frequency.
- Heads: Pivot and follow some detectable phenonomena, such as a light (These are popular in the BEAM community. They can be stand-alone robots, but are more often incorporated into a larger robot.).
- Vibrators: Use a small pager motor with an off-center weight to shake themselves about.

SLIDER (BEAM)

In BEAM robotics, a **Slider** is a robot that has a mode of locomotion by moving body parts smoothly along a surface while remaining in contact with it.

Applications and Current Progress

At present, autonomous robots have seen limited commercial application, with some exceptions such as the iRobot Roomba robotic vacuum cleaner and a few lawn-mowing robots. The main practical application of BEAM has been in the rapid prototyping of motion systems and hobby/education applications. Mark Tilden has successfully used BEAM for the prototyping of products for Wow-Wee Robotics, as evidenced by B.I.O.Bug and RoboRaptor. Solarbotics Ltd., Bug'n'Bots, JCM InVentures Inc., and PagerMotors.com have also brought BEAM-related hobby and educational goods to the marketplace. Vex has also developed Hexbugs, tiny BEAM robots.

Aspiring BEAM roboticists often have problems with the lack of direct control over "pure" BEAM control circuits. There is ongoing work to evaluate biomorphic techniques that copy natural systems because they seem to have an incredible performance advantage over traditional techniques. There are many examples of how tiny insect brains are capable of far better performance than the most advanced microelectronics.

Another barrier to widespread application of BEAM technology is the perceived random nature of the 'nervous network', which requires new techniques to be learned by the builder to successfully diagnose and manipulate the characteristics of the circuitry. A think-tank of international academics meet annually in Telluride, Colorado to address this issue directly, and until recently, Mark Tilden has been part of this effort (he had to withdraw due to his new commercial commitments with Wow-Wee toys).

Having no long-term memory, BEAM robots generally do not learn from past behaviour. However, there has been work in the BEAM community to address this issue. One of the most advanced BEAM robots in this vein is Bruce Robinson's Hider, which has an impressive degree of capability for a microprocessor-less design.

Publications

Patents

- U.S. Patent 613,809 - *Method of and Apparatus for Controlling Mechanism of Moving Vehicle or Vehicles* - Tesla's "telautomaton» patent; First logic gate.
- U.S. Patent 5,325,031 - *Adaptive robotic nervous systems and control circuits therefor* - Tilden's patent; A self-stabilizing control circuit utilizing pulse delay circuits for controlling the limbs of a limbed robot, and a robot incorporating such a circuit; artificial "neurons".

REFERENCES

1. Conrad, James M., and Jonathan W. Mills, "*Stiquito: advanced experiments with a simple and inexpensive robot*", *The future for nitinol-propelled walking robots*, Mark W. Tilden. Los Alamitos, Calif., IEEE Computer Society Press, c1998. LCCN 96029883 ISBN 0-8186-7408-3
2. Tilden, Mark W., and Brosl Hasslacher, "*Living Machines*". Los Alamos National Laboratory, Los Alamos, NM 87545, USA.
3. Tilden, Mark W. and Brosl Hasslacher, "*The Design of "Living" Biomech Machines: How low can one go?"*". Los Alamos National Laboratory, Los Alamos, NM 87545, USA.
4. Still, Susanne, and Mark W. Tilden, "*Controller for a four legged walking machine*". ETH Zuerich, Institute of Neuroinformatics, and Biophysics Division, Los Alamos National Laboratory.
5. Braitenberg, Valentino, "*Vehicles: Experiments in Synthetic Psychology*", 1984. ISBN 0-262-52112-1
6. Rietman, Ed, "*Experiments In Artificial Neural Networks*", 1988. ISBN 0-8306-0237-2
7. Tilden, Mark W., and Brosl Hasslacher, "*Robotics and Autonomous Machines: The Biology and Technology of Intelligent Autonomous Agents*", LANL Paper ID: LA-UR-94-2636, Spring 1995.
8. Dewdney, A.K. "*Photovores: Intelligent Robots are Constructed From Castoffs*". Scientific American Sept 1992, v267, n3, p42(1)
9. Smit, Michael C., and Mark Tilden, "*Beam Robotics*". Algorithm, Vol. 2, No. 2, March 1991, Pg 15-19.
10. Hrynkiw, David M., and Tilden, Mark W., "*Junkbots, Bugbots, and Bots on Wheels*", 2002. ISBN 0-07-222601-3 (Book support website)

CHAPTER 8

Nanorobotics

Chapter Contents

Nanorobots are theoretical microscopic devices measured on the scale of nanometers (1nm equals one millionth of 1 millimeter). When fully realized from the hypothetical stage, they would work at the atomic, molecular and cellular level to perform tasks in both the medical and industrial fields that have heretofore been the stuff of science fiction.

INTRODUCTION

Nanorobotics is an emerging technology field creating machines or robots which components are at or near the scale of a nanometre (10^{-9} meters). More specifically, nanorobotics (as opposed to microrobotics) refers to the nanotechnology engineering discipline of designing and building nanorobots, with devices ranging in size from 0.1–10 micrometres and constructed of nanoscale or molecular components. The terms *nanobot*, *nanoid*, *nanite*, *nanomachine*, or *nanomite* have also been used to describe such devices currently under research and development.

Nanomachines are largely in the research and development phase, but some primitive molecular machines and nanomotors have been tested. An example is a sensor having a switch approximately 1.5 nanometers across, able to count specific molecules in a chemical sample. The first useful applications of nanomachines may be in nanomedicine. For example, biological machines could be used to identify and destroy cancer cells. Another potential application is the detection of toxic chemicals, and the measurement of their concentrations, in the environment. Rice University has demonstrated a single-molecule car developed by a chemical process and including Buckminsterfullerenes (buckyballs) for wheels. It is actuated by controlling the environmental temperature and by positioning a scanning tunneling microscope tip.

Another definition is a robot that allows precise interactions with nanoscale objects, or can manipulate with nanoscale resolution. Such devices are more related to microscopy or scanning probe microscopy, instead of the description of nanorobots as molecular machine. Using the microscopy definition, even a large apparatus such as an atomic force microscope can be considered a nanorobotic instrument when configured to perform nanomanipulation. For this viewpoint, macroscale robots or microrobots that can move with nanoscale precision can also be considered nanorobots.

NANOROBOTICS THEORY

According to Richard Feynman, it was his former graduate student and collaborator Albert Hibbs who originally suggested to him (circa 1959) the idea of a *medical* use for Feynman's theoretical micromachines (see nanotechnology). Hibbs suggested that certain repair machines might one day be reduced in size to the point that it would, in theory, be possible to (as Feynman put it) *swallow the surgeon*". The idea was incorporated into Feynman's 1959 essay *There's Plenty of Room at the Bottom.*

Since nanorobots would be microscopic in size, it would probably be necessary for very large numbers of them to work together to perform microscopic and macroscopic tasks. These nanorobot swarms, both those unable to replicate (as in utility fog) and those able to replicate unconstrainedly in the natural environment (as in grey goo and its less common variants, such as synthetic biology or utility fog), are found in many science fiction stories, such as the Borg nanoprobes in *Star Trek* and The Outer Limits episode The New Breed.

Some proponents of nanorobotics, in reaction to the grey goo scenarios that they earlier helped to propagate, hold the view that nanorobots able to replicate outside of a restricted factory environment do not form a necessary part of a purported productive nanotechnology, and that the process of self-replication, were it ever to be developed, could be made inherently safe. They further assert that their current plans for developing and using molecular manufacturing do not in fact include free-foraging replicators.

The most detailed theoretical discussion of nanorobotics, including specific design issues such as sensing, power communication, navigation, manipulation, locomotion, and onboard computation, has been presented in the medical context of nanomedicine by Robert Freitas. Some of these discussions remain at the level of unbuildable generality and do not approach the level of detailed engineering.

BIOCHIP

In molecular biology, biochips are essentially miniaturized laboratories that can perform hundreds or thousands of simultaneous biochemical reactions. Biochips enable researchers to quickly screen large numbers of biological analytes for a variety of purposes, from disease diagnosis to detection of bioterrorism agents. Digital microfluidic biochips have become one of the most promising technologies in many biomedical fields. In a digital microfluidic biochip, a group of (adjacent) cells in the microfluidic array can be configured to work as storage, functional operations, as well as for transporting fluid droplets dynamically.

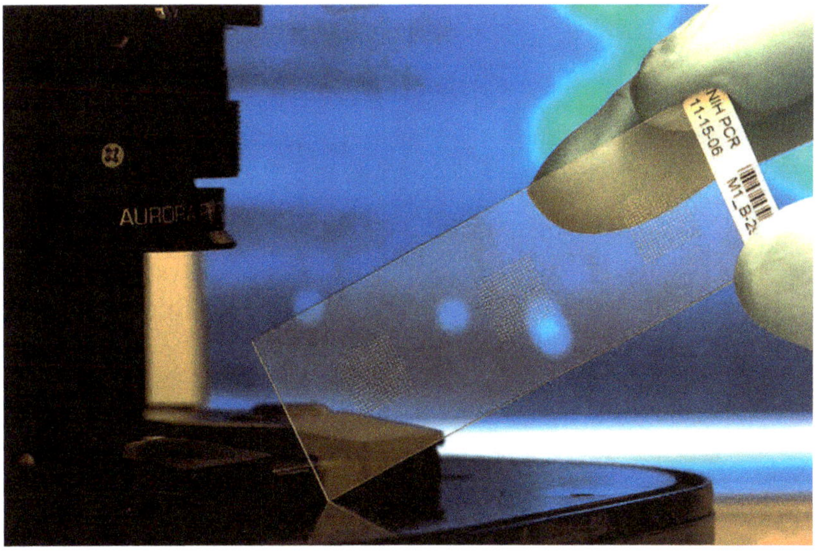

Hundreds of gel drops are visible on the biochip.

The development started with early work on the underlying sensor technology. One of the first portable, chemistry-based sensors was the glass pH electrode, invented in 1922 by Hughes. In subsequent years. For example, a K^+ sensor was produced by incorporating valinomycin into a thin membrane.

In 1953, Watson and Crick announced their discovery of the now familiar double helix structure of DNA molecules and set the stage for genetics research that continues to the present day. The development of sequencing techniques in 1977 by Gilbert and Sanger (working separately) enabled researchers to directly read the genetic codes that provide instructions for protein synthesis. This research showed how hybridization of complementary single oligonucleotide strands could be used as a basis for DNA sensing. Two additional developments enabled the technology used in modern DNA-based. First, in 1983 Kary Mullis invented the polymerase chain reaction (PCR) technique, a method for amplifying DNA concentrations. This discovery made possible the detection of extremely small quantities of DNA in samples. Secondly in 1986 Hood and co-workers devised a method to label DNA molecules with fluorescent tags instead of radiolabels, thus enabling hybridization experiments to be observed optically.

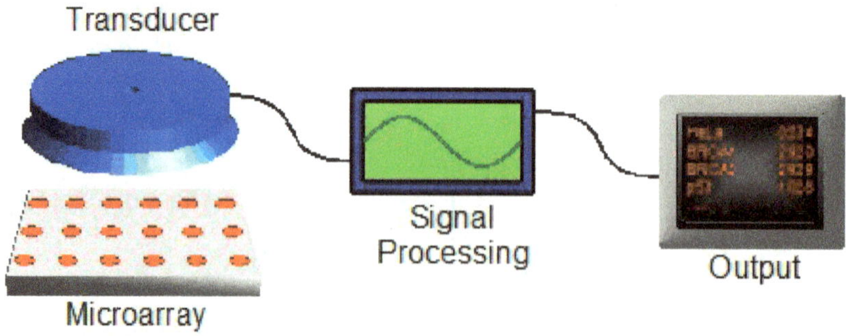

Figure 1. Biochips are a platform that require, in addition to microarray technology, transduction and signal processing technologies to output the results of sensing experiments.

Figure 1 shows the make up of a typical biochip platform. The actual sensing component (or "chip") is just one piece of a complete analysis system Transduction must be done to translate the actual sensing event (DNA binding, oxidation/reduction, *etc.*) into a format understandable by a computer (voltage, light intensity, mass, *etc.*), which then enables additional analysis and processing to produce a final, human-readable output. The multiple technologies needed to make a successful biochip — from sensing chemistry, to microarraying, to signal processing — require a true multidisciplinary approach, making the barrier to entry steep. One of the first commercial biochips was introduced by Affymetrix. Their "GeneChip" products contain thousands of individual DNA sensors for use in sensing defects, or single nucleotide polymorphisms (SNPs), in genes such as p53 (a tumor suppressor) and BRCA1 and BRCA2 (related to breast cancer). The chips are produced using microlithography techniques traditionally used to fabricate integrated circuits .

Microarray Fabrication

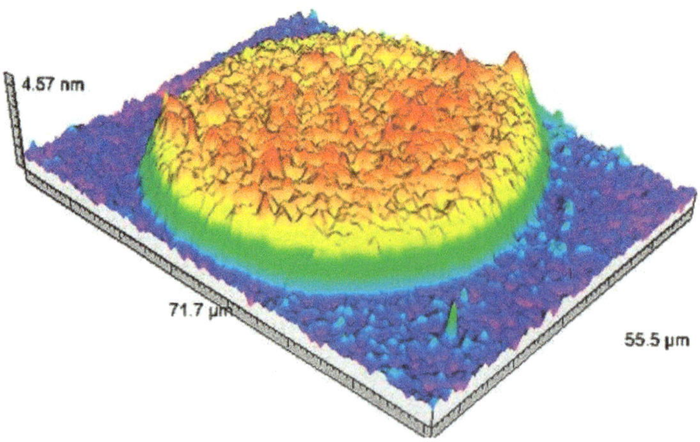

3D Sarfus image of a DNA biochip.

The microarray — the dense, two-dimensional grid of biosensors — is the critical component of a biochip platform. Typically, the sensors are deposited on a flat substrate, which may either be passive (*e.g.* silicon or glass) or active, the latter consisting of integrated electronics or micromechanical devices that perform or assist signal transduction. Surface chemistry is used to covalently bind the sensor molecules to the substrate medium. The fabrication of microarrays is non-trivial and is a major economic and technological hurdle that may ultimately decide the success of future biochip platforms. The primary manufacturing challenge is the process of placing each sensor at a specific position (typically on a Cartesian grid) on the substrate. Various means exist to achieve the placement, but typically robotic micro-pipettingor micro-printing systems are used to place tiny spots of sensor material on the chip surface. Because each sensor is unique, only a few spots can be placed at a time. The low-throughput nature of this process results in high manufacturing costs.

Fodor and colleagues developed a unique fabrication process (later used by Affymetrix) in which a series of microlithography steps is used to combinatorially synthesize hundreds of thousands of unique, single-stranded DNA sensors on a substrate one nucleotide at a time. One lithography step is needed per base type; thus, a total of four steps is required per nucleotide level. Although this technique is very powerful in that many sensors can be created simultaneously, it is currently only feasible for creating short DNA strands (15–25 nucleotides). Reliability and cost factors limit the number of photolithography steps that can be done. Furthermore, light-directed combinatorial synthesis techniques are not currently possible for proteins or other sensing molecules.

As noted above, most microarrays consist of a Cartesian grid of sensors. This approach is used chiefly to map or encode the coordinate of each sensor to its function. Sensors in these arrays typically use a universal signalling technique (*e.g.* fluorescence), thus making coordinates their only identifying feature. These arrays must be made using a serial process (*i.e.* requiring multiple, sequential steps) to ensure that each sensor is placed at the correct position.

Random fabrication, in which the sensors are placed at arbitrary positions on the chip, is an alternative to the serial method. The tedious and expensive positioning process is not required, enabling the use of parallelized self-assembly techniques. In this approach, large batches of identical sensors can be produced; sensors from each batch are then combined and assembled into an array. A non-coordinate based encoding scheme must be used to identify each sensor. As the figure shows, such a design was first demonstrated (and later commercialized by Illumina) using functionalized beads placed randomly in the wells of an etched fiber optic cable. Each bead was uniquely encoded with a fluorescent signature. However, this encoding scheme is limited in the number of unique dye combinations that can be used and successfully differentiated.

Protein Biochip Array and Other Microarray Technologies

Microarrays are not limited to DNA analysis; protein microarrays, antibody microarray, chemical compound microarray can also be produced using biochips. Randox Laboratories Ltd. launched Evidence, the first protein Biochip Array Technology analyzer in 2003. In protein Biochip Array Technology, the biochip replaces the ELISA plate or cuvette as the reaction platform. The biochip is used to simultaneously analyze a panel of related tests in a single sample, producing a patient profile. The patient profile can be used in disease screening, diagnosis, monitoring disease progression or monitoring treatment. Performing multiple analyses simultaneously, described as multiplexing, allows a significant reduction in processing time and the amount of patient sample required. Biochip Array Technology is a novel application of a familiar methodology, using sandwich, competitive and antibody-capture immunoassays. The difference from conventional immunoassays is that, the capture ligands are covalently attached to the surface of the biochip in an ordered array rather than in solution.

In sandwich assays an enzyme-labelled antibody is used; in competitive assays an enzyme-labelled antigen is used. On antibody-antigen binding a chemiluminescence reaction produces light. Detection is by a charge-coupled device (CCD) camera. The CCD camera is a sensitive and high-resolution sensor able to accurately detect and quantify very low levels of light. The test regions are located using a grid pattern then the chemiluminescence signals are analysed by imaging software to rapidly and simultaneously quantify the individual analytes.

Details about other array technologies can be found in the following page: Antibody microarray

Nubots

A *nucleic acid robot* (nubot) is an organic molecular machine at the nanoscale. DNA structure can provide means to assemble 2D and 3D nanomechanical devices. DNA based machines can be activated using small molecules, proteins and other molecules of DNA. Biological circuit gates based on DNA materials have been engineered as molecular machines to allow in-vitro drug delivery for targeted health problems.Such material based systems would work most closely to smart biomaterial drug system delivery, while not allowing precise in vivo teleoperation of such engineered prototypes.

DNA MACHINE

A **DNA machine** is a molecular machine constructed from DNA. Research into DNA machines was pioneered in the late 1980s by Nadrian Seeman and co-workers from New York University. DNA is used because of the numerous biological tools already found in nature that can affect DNA, and the immense knowledge of how DNA works previously researched by biochemists.

DNA machines can be logically designed since DNA assembly of the double helix is based on strict rules of base pairing that allow portions of the strand to be predictably connected based on their sequence. This selective stickiness is a key advantage in the construction of DNA machines.

An example of a DNA machine was reported by Bernard Yurke and co-workers at Lucent Technologies in the year 2000, who constructed molecular tweezers out of DNA. The DNA tweezers contain three strands: A, B and C. Strand A latches onto half of strand B and half of strand C, and so it joins them all together. Strand A acts as a hinge so that the two arms — AB and AC — can move. The structure floats with its arms open wide. They can be pulled shut by adding a fourth strand of DNA (D) programmed to stick to both of the dangling, unpaired sections of strands B and C. The closing of the tweezers was proven by tagging strand A at either end with light-emitting molecules that do not emit light when they are close together. To re-open the tweezers add a further strand (E) with the right sequence to pair up with strand D. Once paired up, they have no connection to the machine BAC, so float away. The DNA machine can be opened and closed repeatedly by cycling between strands D and E. These tweezers can be used for removing drugs from inside fullerenes as well as from a self assembled DNA tetrahedron. The state of the device can be determined by measuring the separation between donor and acceptor fluorophores using FRET.

DNA walkers are another type of DNA machine.

Surface-bound systems

Several reports have demonstrated the attachment of synthetic molecular motors to surfaces. These primitive nanomachines have been shown to undergo machine-like motions when confined to the surface of a macroscopic material. The surface anchored motors could potentially be used to move and position nanoscale materials on a surface in the manner of a conveyor belt.

Positional Nanoassembly

Nanofactory Collaboration, founded by Robert Freitas and Ralph Merkle in 2000 and involving 23 researchers from 10 organizations and 4 countries, focuses on developing a practical research agenda specifically aimed at developing positionally-controlled diamond mechanosynthesis and a diamondoid nanofactory that would have the capability of building diamondoid medical nanorobots.

Bacteria-Based

This approach proposes the use of biological microorganisms, like the bacterium *Escherichia coli* and *Salmonella typhimurium*. Thus the model uses a flagellum for propulsion purposes. Electromagnetic fields normally control the motion of this kind of biological integrated device.Chemists at the University of Nebraska have created a humidity gauge by fusing a bacterium to a silicone computer chip.

Virus-Based

Retroviruses can be retrained to attach to cells and replace DNA. They go through a process called reverse transcription to deliver genetic packaging in a vector. Usually, these devices are Pol – Gag genes of the virus for the Capsid and Delivery system. This process is called retroviral gene therapy, having the ability to re-engineer cellular DNA by usage of viral vectors. This approach has appeared in the form of retroviral, adenoviral, and lentiviral gene delivery systems. These gene therapy vectors have been used in cats to send genes into the genetically modified organism (GMO), causing it to display the trait.

Open Technology

A document with a proposal on nanobiotech development using open design technology methods, as in open-source hardware and open-source software, has been addressed to the United Nations General Assembly. According to the document sent to the United Nations, in the same way that open source has in recent years accelerated the development of computer systems, a similar approach should benefit the society at large and accelerate nanorobotics development. The use of nanobiotechnology should be established as a human heritage for the coming generations, and developed as an open technology based on ethical practices for peaceful purposes. Open technology is stated as a fundamental key for such an aim.

Nanorobot Race

In the same ways that technology research and development drove the space race and nuclear arms race, a race for nanorobots is occurring. There is plenty of ground allowing nanorobots to be included among the emerging technologies. Some of the reasons are that large corporations, such as General Electric, Hewlett-Packard, Synopsys, Northrop Grumman and Siemens have been recently working in the development and research of nanorobots; surgeons are getting involved and starting to propose ways to apply nanorobots for common medical procedures; universities and research institutes were granted funds by government agencies exceeding $2 billion towards research developing nanodevices for medicine; bankers are also strategically investing with the intent to acquire beforehand rights and royalties on future nanorobots commercialisation. Some aspects of nanorobot litigation and related issues linked to monopoly have already arisen. A large number of patents has been granted recently on nanorobots, done mostly for patent agents, companies specialized solely on building patent portfolios, and lawyers. After a long series of patents and eventually litigations, see for example the Invention of Radio, or the War of Currents, emerging fields of technology tend to become a monopoly, which normally is dominated by large corporations.

3D PRINTING

3D printing, also known as **additive manufacturing** (**AM**), refers to processes used to synthesize a three-dimensional object in which successive layers of material are formed under computer control to create an object. Objects can be of almost any shape or geometry and are produced using digital model data from a 3D model or another electronic data source such as an Additive Manufacturing File (AMF) file.

Futurologist Jeremy Rifkin claimed that 3D printing or AM signals the beginning of a third industrial revolution, succeeding the production line assembly that dominated manufacturing starting in the late 19th century.

The term 3D printing's origin sense is in reference to a process that deposits a binder material onto a powder bed with inkjet printer heads layer by layer. More recently, the term is being used in popular vernacular to encompass a wider variety of additive manufacturing techniques. United States and global Technical standards use the official term *additive manufacturing* for this broader sense. ISO/ASTM52900-15 defines seven categories of AM processes within its meaning: Binder Jetting, Directed Energy Deposition, Material Extrusion, Material Jetting, Powder Bed Fusion, Sheet Lamination and Vat Photopolymerization.

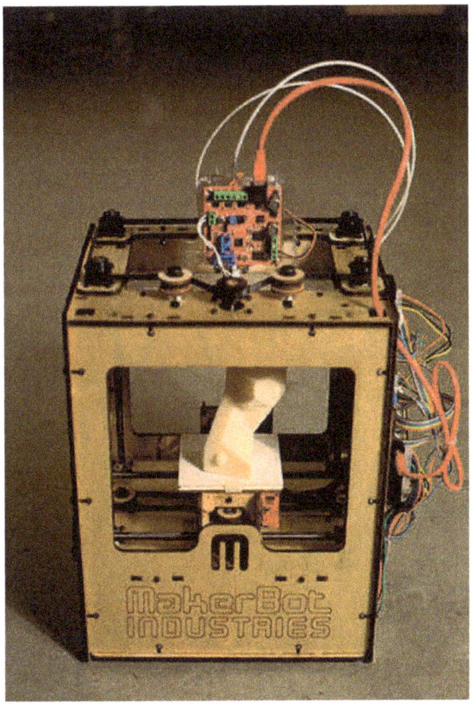

A MakerBot three-dimensional printer.

Terminology and Methods

Early additive manufacturing equipment and materials were developed in the 1980s. In 1981, Hideo Kodama of Nagoya Municipal Industrial Research Institute invented two AM fabricating methods of a three-dimensional plastic model with photo-hardening polymer, where the UV exposure area is controlled by a mask pattern or the scanning fiber transmitter. But on July 16, 1984 Alain Le Méhauté, Olivier de Witte and Jean Claude André filed their patent for the stereolithography process. It was three weeks before Chuck Hull filed his own patent for stereolithography. The application of French inventors were abandoned by the French General Electric Company (now Alcatel-Alsthom) and CILAS (The Laser Consortium). The claimed reason was "for lack of business perspective". Then in 1984, Chuck Hull of 3D Systems Corporationdeveloped a prototype system based on a process known as stereolithography, in which layers are added by curing photopolymers with ultraviolet light lasers. Hull defined the process as a "system for generating three-dimensional objects by creating a cross-sectional pattern of the object to be formed," but this had been already invented by Kodama. Hull's contribution is the design of the STL (STereoLithography) file format widely accepted by 3D printing software as well as the digital slicing and infill strategies common to many processes today. The term *3D printing* originally referred to a process employing standard and custom inkjet print heads. The technology used by most 3D printers to date—especially hobbyist and consumer-oriented models—is fused deposition modeling, a special application of plastic extrusion.

AM processes for metal sintering or melting (such as selective laser sintering, direct metal laser sintering, and selective laser melting) usually went by their own individual names in the 1980s and 1990s. At the time, nearly all metal working was produced by casting, fabrication, stamping, and machining; although plenty of automation was applied to those technologies (such as by robot welding and CNC), the idea of a tool or head moving through a 3D work envelope transforming a mass of raw material into a desired shape layer by layer was associated by most people only with processes that removed metal (rather than adding it), such as CNC milling, CNC EDM, and many others. But AM-type sintering was beginning to challenge that assumption. By the mid 1990s, new techniques for material deposition were developed at Stanford and Carnegie Mellon University, including microcasting and sprayed materials. Sacrificial and support materials had also become more common, enabling new object geometries.

The umbrella term *additive manufacturing* gained wider currency in the decade of the 2000s. As the various additive processes matured, it became clear that soon metal removal would no longer be the only metalworking process done under that type of control (a tool or head moving through a 3D work envelope transforming a mass of raw material into a desired shape layer by layer). It was during this decade that the term *subtractive manufacturing* appeared as a retronym for the large family of machining processes with metal removal as their common theme. At this time, the term *3D printing* still referred only to the polymer technologies in most minds, and the term *AM* was likelier to be used in metalworking and end use part production contexts than among polymer/inkjet/stereolithography enthusiasts. The term *subtractive* has not replaced the term *machining*, instead complementing it when a term that covers any removal method is needed.

By the early 2010s, the terms *3D printing* and *additive manufacturing* evolved senses in which they were alternate umbrella terms for AM technologies, one being used in popular vernacular by consumer - maker communities and the media, and the other used officially by industrial AM end use part producers, AM machine manufacturers, and global technical standards organizations.

Both terms reflect the simple fact that the technologies all share the common theme of sequential-layer material addition/joining throughout a 3D work envelope under automated control.

(Other terms that had been used as AM synonyms (although sometimes as hypernyms), included *desktop manufacturing*, *rapid manufacturing, agile tooling* as the logical production-level successor to *rapid prototyping*, and *on-demand manufacturing* which echoes *on-demand printing* in the 2D sense of *printing*.) The 2010s were the first decade in which metal end use parts such as engine brackets and large nuts would be grown (either before or instead of machining) in job production rather than obligately being machined from bar stock or plate.

Agile tooling is a term used to describe the process of using modular means to design tooling that is produced by additive manufacturing or 3D printing methods to enable quick prototyping and responses to tooling and fixture needs. Agile tooling uses a cost effective and high quality method to quickly respond to customer and market needs. It can be used in hydro-forming, stamping, injection molding and other manufacturing processes.

As technology matured, several authors had begun to speculate that 3D printing could aid in sustainable development in the developing world.

General Principles

Modelling

3D printable models may be created with a computer-aided design (CAD) package, via a 3D scanner, or by a plain digital camera and photogrammetry software. 3D printed models created with CAD result in reduced errors and can be corrected before printing, allowing verification in the design of the object before it is printed.

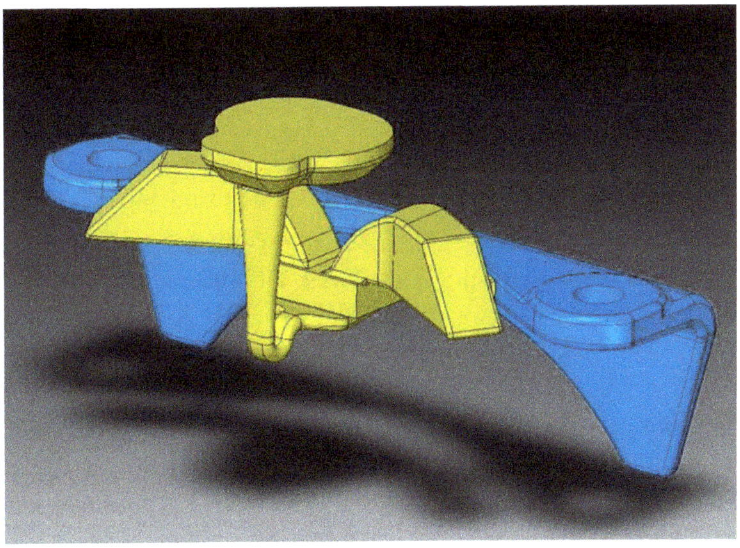

CAD model used for 3D printing

The manual modeling process of preparing geometric data for 3D computer graphics is similar to plastic arts such as sculpting. 3D scanning is a process of collecting digital data on the shape and appearance of a real object, creating a digital model based on it.

Printing

Before printing a 3D model from an STL file, it must first be examined for errors. Most CAD applications produce errors in output STL files: holes, faces normals, self-intersections, noise shells or manifold errors. A step in the STL generation known as repair fixes such problems in the original model. Generally STLs that have been produced from a model obtained through 3D scanning often have more of these errors. This is due to how 3D scanning works-as it is often by point to point acquisition, reconstruction will include errors in most cases.

Timelapse video of a hyperboloid object (designed by George W. Hart) made of PLA using a RepRap "Prusa Mendel 3 printer for molten polymer deposition

Once completed, the STL file needs to be processed by a piece of software called a slicer, which converts the model into a series of thin layers and produces a G-code file containing instructions tailored to a specific type of 3D printer (FDM printers). This G-code file can then be printed with 3D printing client software (which loads the G-code, and uses it to instruct the 3D printer during the 3D printing process).

Printer resolution describes layer thickness and X-Y resolution in dots per inch (dpi) or micrometers (μm). Typical layer thickness is around 100 μm (250 DPI), although some machines can print layers as thin as 16 μm (1,600 DPI). X-Y resolution is comparable to that of laser printers. The particles (3D dots) are around 50 to 100 μm (510 to 250 DPI) in diameter.

Construction of a model with contemporary methods can take anywhere from several hours to several days, depending on the method used and the size and complexity of the model. Additive systems can typically reduce this time to a few hours, although it varies widely depending on the type of machine used and the size and number of models being produced simultaneously.

Traditional techniques like injection moulding can be less expensive for manufacturing polymer products in high quantities, but additive manufacturing can be faster, more flexible and less expensive when producing relatively small quantities of parts. 3D printers give designers and concept development teams the ability to produce parts and concept models using a desktop size printer.

Seemingly paradoxically, more complex objects can be cheaper for 3D printing production than less complex objects.

Finishing

Though the printer-produced resolution is sufficient for many applications, printing a slightly oversized version of the desired object in standard resolution and then removing material with a higher-resolution subtractive process can achieve greater precision.

Some printable polymers such as ABS, allow the surface finish to be smoothed and improved using chemical vapor processes based on acetone or similar solvents.

Some additive manufacturing techniques are capable of using multiple materials in the course of constructing parts. These techniques are able to print in multiple colors and color combinations simultaneously, and would not necessarily require painting.

Some printing techniques require internal supports to be built for overhanging features during construction. These supports must be mechanically removed or dissolved upon completion of the print.

All of the commercialized metal 3D printers involve cutting the metal component off the metal substrate after deposition. A new process for the GMAW 3D printing allows for substrate surface modifications to remove aluminum or steel.

Processes

Several 3D printing processes have been invented since the late 1970s. The printers were originally large, expensive, and highly limited in what they could produce.

A large number of additive processes are now available. The main differences between processes are in the way layers are deposited to create parts and in the materials that are used. Some methods melt or soften the material to produce the layers, for example. selective laser melting (SLM) or direct metal laser sintering (DMLS), selective laser sintering (SLS), fused deposition modeling (FDM), or fused filament fabrication (FFF), while others cure liquid materials using different sophisticated technologies, such as stereolithography (SLA). With laminated object manufacturing (LOM), thin layers are cut to shape and joined together (e.g., paper, polymer, metal). Each method has its own advantages and drawbacks, which is why some companies offer a choice of powder and polymer for the material used to build the object. Others sometimes use standard, off-the-shelf business paper as the build material to produce a durable prototype. The main considerations in choosing a machine are generally speed, costs of the 3D printer, of the printed prototype, choice and cost of the materials, and color capabilities.

Printers that work directly with metals are generally expensive. However less expensive printers can be used to make a mold, which is then used to make metal parts.

Type	Technologies	Materials
Extrusion	Fused deposition modeling (FDM) or Fused filament fabrication (FFF)	Thermoplastics, eutectic metals, edible materials, Rubbers, Modeling clay, Plasticine, Metal clay (including Precious Metal Clay)
	Robocasting or Direct Ink Writing (DIW)	Ceramic materials, Metal alloy, cermet, metal matrix composite, ceramic matrix composite
Light polymerized	Stereolithography (SLA)	Photopolymer
	Digital Light Processing (DLP)	Photopolymer

Powder Bed	Powder bed and inkjet head 3D printing (3DP)	Almost any metal alloy, powdered polymers, Plaster
	Electron-beam melting (EBM)	Almost any metal alloy including Titanium alloys
	Selective laser melting (SLM)	Titanium alloys, Cobalt Chrome alloys, Stainless Steel, Aluminium
	Selective heat sintering (SHS)	Thermoplastic powder
	Selective laser sintering (SLS)	Thermoplastics, metal powders, ceramic powders
	Direct metal laser sintering (DMLS)	Almost any metal alloy
Laminated	Laminated object manufacturing (LOM)	Paper, metal foil, plastic film
Powder fed	Directed Energy Deposition	Almost any metal alloy
Wire	Electron beam freeform fabrication (EBF^3)	Almost any metal alloy

Extrusion Deposition

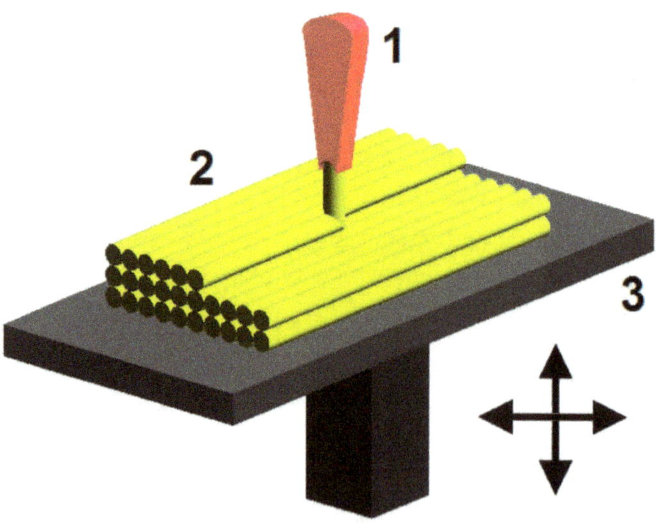

Fused deposition modeling: 1 – nozzle ejecting molten material, 2 – deposited material (modeled part), 3 – controlled movable table

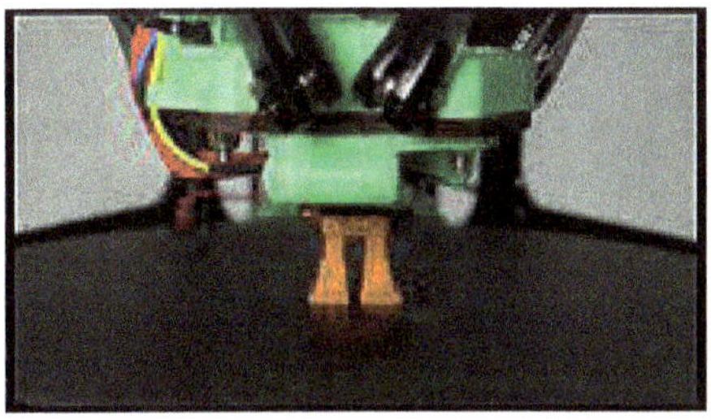

A timelapse video of a robot model (logo of Make magazine) being printed using FDM on a RepRapPro Fisher printer.

Fused deposition modeling (FDM), derives from automatic polymeric foil hot air welding system, hot-melt gluing and automatic gasket deposition. Such principle has been further developed by S. Scott Crump in the late 1980s and was commercialized in 1990 by Stratasys. After the patent on this technology expired, a large open-source development community developed and both commercial and DIY variants utilizing this type of 3D printer appeared. As a result, the price of this technology has dropped by two orders of magnitude since its creation.

In fused deposition modeling the model or part is produced by extruding small beads of material which harden immediately to form layers. A thermoplastic filament or metal wire that is wound on a coil is unreeled to supply material to an extrusion nozzle head (3D printer extruder). The nozzle head heats the material and turns the flow on and off. Typically stepper motors or servo motors are employed to move the extrusion head and adjust the flow. The printer usually has 3 axes of motion. A computer-aided manufacturing (CAM) software package is used to generate the G-Code that is sent to a microcontroller which controls the motors.

Extrusion in 3D printing using material extrusion involves a cold end and a hot end.

Various polymers are used, including acrylonitrile butadiene styrene (ABS), polycarbonate (PC), polylactic acid (PLA), high-density polyethylene (HDPE), PC/ABS, polyphenylsulfone (PPSU) and high impact polystyrene (HIPS). In general, the polymer is in the form of a filament fabricated from virgin resins. There are multiple projects in the open-sourced community aimed at processing post-consumer plastic waste into filament. These involve machines used to shred and extrude the plastic material into filament. Additionally, fluoropolymers such as PTFE tubing are used in the process due to the material's ability to withstand high temperatures. This ability is especially useful in transferring filaments.

FDM is somewhat restricted in the variation of shapes that may be fabricated. For example, FDM usually cannot produce stalactite-like structures, since they would be unsupported during the build. Otherwise, a thin support must be designed into the structure, which can be broken away during finishing. Fused deposition modeling is also referred to as fused filament fabrication (FFF) by companies who do not hold the original patents like Stratasys does.

Binding of Granular Materials

Another 3D printing approach is the selective fusing of materials in a granular bed. The technique fuses parts of the layer and then moves upward in the working area, adding another layer of granules and repeating the process until the piece has built up. This process uses the unfused media to support overhangs and thin walls in the part being produced, which reduces the need for temporary auxiliary supports for the piece. A laser is typically used to sinter the media into a solid. Examples include selective laser sintering (SLS), with both metals and polymers (e.g., PA, PA-GF, Rigid GF, PEEK, PS, Alumide, Carbonmide, elastomers), and direct metal laser sintering (DMLS).

Selective Laser Sintering (SLS) was developed and patented by Dr. Carl Deckard and Dr. Joseph Beaman at the University of Texas at Austin in the mid-1980s, under sponsorship of DARPA. A similar process was patented without being commercialized by R. F. Housholder in 1979.

Selective laser melting (SLM) does not use sintering for the fusion of powder granules but will completely melt the powder using a high-energy laser to create fully dense materials in a layer-wise method that has mechanical properties similar to those of conventional manufactured metals.

Electron beam melting (EBM) is a similar type of additive manufacturing technology for metal parts (e.g. titanium alloys). EBM manufactures parts by melting metal powder layer by layer with an electron beam in a high vacuum. Unlike metal sintering techniques that operate below melting point, EBM parts are void-free.

Another method consists of an inkjet 3D printing system. The printer creates the model one layer at a time by spreading a layer of powder (plaster, or resins) and printing a binder in the cross-section of the part using an inkjet-like process. This is repeated until every layer has been printed. This technology allows the printing of full color prototypes, overhangs, and elastomer parts. The strength of bonded powder prints can be enhanced with wax or thermoset polymer impregnation.

Lamination

In some printers, paper can be used as the build material, resulting in a lower cost to print. During the 1990s some companies marketed printers that cut cross sections out of special adhesive coated paper using a carbon dioxide laser and then laminated them together.

In 2005 Mcor Technologies Ltd developed a different process using ordinary sheets of office paper, a tungsten carbide blade to cut the shape, and selective deposition of adhesive and pressure to bond the prototype. There are also a number of companies selling printers that print laminated objects using thin plastic and metal sheets.

Photopolymerization

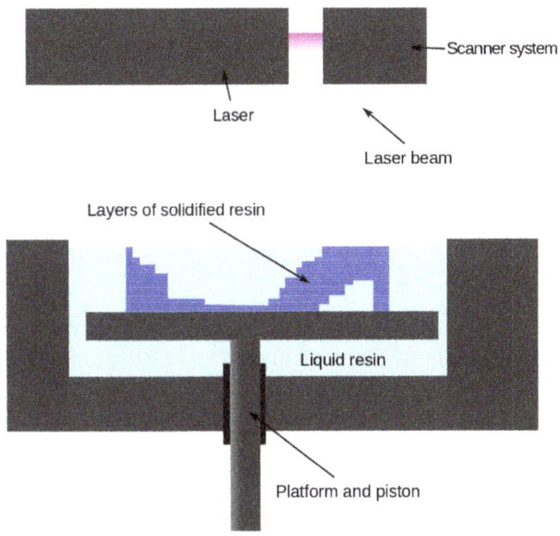

Stereolithography apparatus

Stereolithography was patented in 1986 by Chuck Hull. Photopolymerization is primarily used in stereolithography (SLA) to produce a solid part from a liquid. This process was a dramatic departure from the photosculpture method of François Willème (1830–1905) developed in 1860 and the photopolymerization of Mitsubishi's Matsubara in 1974.

The photosculpture method consisted of photographing a subject from a variety of equidistant angles and projecting each photograph onto a screen, where a pantograph was used to trace the outline onto modeling clay. In photo-polymerization, a vat of liquid polymer is exposed to controlled lighting under safelight conditions. The exposed liquid polymer hardens. Polymerization occurs when photopolymers are exposed to light when photopolymers contain chromophores, otherwise, the addition of molecules that are photosensitive are utilized to react with the solution to begin polymerization. Polymerization of monomers lead to cross-linking, which creates a polymer. Through these covalent bonds, the property of the solution is changed. The build plate then moves down in small increments and the liquid polymer is again exposed to light. The process repeats until the model has been built. The liquid polymer is then drained from the vat, leaving the solid model. The EnvisionTEC*Perfactory* is an example of a DLP rapid prototyping system.

Inkjet printer systems like the *Objet PolyJet* system spray photopolymer materials onto a build tray in ultra-thin layers (between 16 and 30 μm) until the part is completed. Each photopolymer layer is cured with UV light after it is jetted, producing fully cured models that can be handled and used immediately, without post-curing. The gel-like support material, which is designed to support complicated geometries, is removed by hand and water jetting. It is also suitable for elastomers.

Ultra-small features can be made with the 3D micro-fabrication technique used in multiphoton photopolymerisation. This approach uses a focused laser to trace the desired 3D object into a block of gel. Due to the nonlinear nature of photo excitation, the gel is cured to a solid only in the places where the laser was focused while the remaining gel is then washed away. Feature sizes of under 100 nm are easily produced, as well as complex structures with moving and interlocked parts.

Yet another approach uses a synthetic resin that is solidified using LEDs.

In Mask-image-projection-based stereolithography a 3D digital model is sliced by a set of horizontal planes. Each slice is converted into a two-dimensional mask image. The mask image is then projected onto a photocurable liquid resin surface and light is projected onto the resin to cure it in the shape of the layer. The technique has been used to create objects composed of multiple materials that cure at different rates. In research systems, the light is projected from below, allowing the resin to be quickly spread into uniform thin layers, reducing production time from hours to minutes. Commercially available devices such as Objet Connex apply the resin via small nozzles.

Continuous liquid interface production (CLIP) is another form of additive manufacturing that uses the DLP based photo polymerization process to create smooth-sided solid objects of a wide variety of shapes. The continuous process of CLIP begins with a pool of liquid photopolymer resin. Part of the pool bottom is transparent to ultraviolet light (the window). Like DLP systems before it, ultraviolet light beam shines through the window, illuminating the precise cross-section of the object. The light causes the resin to solidify. The object rises slowly enough to allow resin to flow under and maintain contact with the bottom of the object. CLIP is different from traditional DLP processes, due to an oxygen-permeable membrane which lies below the resin, creating a dead zone (persistent liquid interface) preventing the resin from attaching to the window (photopolymerization is inhibited between the window and the polymerizer).

Unlike stereolithography, the printing process is considered continuous by its founders and considerably faster than traditional DLP processes.

Powder Fed Directed Energy Deposition

In powder fed directed energy deposition a high power laser is used to melt metal powder supplied to the focus of the laser beam. The laser beam typically travels through the center of the deposition head and is focused to a small spot by one or more lenses. The build occurs on a X-Y table which is driven by a tool path created from a digital model to fabricate an object layer by layer. The deposition head is moved up vertically as each layer is completed. Metal powder is delivered and distributed around the circumference of the head or can be split by an internal manifold and delivered through nozzles arranged in various configurations around the deposition head. A hermetically sealed chamber filled with inert gas or a local inert shroud gas is often used to shield the melt pool from atmospheric oxygen for better control of material properties. The powder fed directed energy process is similar to Selective Laser Sintering, but the metal powder is applied only where material is being added to the part at that moment. The process supports a wide range of materials including titanium, stainless steel, aluminum, and other specialty materials as well as composites and functionally graded material. The process can not only fully build new metal parts but can also add material to existing parts for example for coatings, repair, and hybrid manufacturing applications. LENS (Laser Engineered Net Shaping), which was developed by Sandia National Labs, is one example of the Powder Fed - Directed Energy Deposition process for 3D printing or restoring metal parts.

Metal Wire Processes

Laser-based wirefeed systems, such as Laser Metal Deposition-wire (LMD-w), feed wire through a nozzle that is melted by a laser using inert gas shielding in either an open environment (gas surrounding the laser), or in a sealed chamber. Electron beam freeform fabrication uses an electron beam heat source inside a vacuum chamber.

It is also possible to use conventional gas metal arc welding attached to a 3D stage to 3-D print metals such as steel and aluminum. Low-cost open source RepRap-style 3-D printers have been outfitted with Arduino-based sensors and demonstrated reasonable metallurgical properties from conventional welding wire as feedstock.

Printers

Industry Use

As of October 2012, additive manufacturing systems were on the market that ranged from $2,000 to $500,000 in price and were employed in industries including aerospace, architecture, automotive, defense, and medical replacements, among many others. For example, General Electric uses the high-end model to build parts for turbines. Many of these systems are used for rapid prototyping, before mass production methods are employed. Higher education has proven to be a major buyer of desktop and professional 3D printers which industry experts generally view as a positive indicator. Significant desktop 3D printer purchases by both K-12 and universities help sustain a desktop 3D printer market that has had problems in 2015-2016, while universities are filling a current talent gap that is expected to bolster the industry in years ahead. Libraries around the world have also become locations to house smaller 3D printers for educational and community access.

Consumer Use

Several projects and companies are making efforts to develop affordable 3D printers for home desktop use. Much of this work has been driven by and targeted at DIY/Maker/enthusiast/early adopter communities, with additional ties to the academic and hacker communities.

RepRap Project is one of the longest running projects in the desktop category. The RepRap project aims to produce a free and open source hardware (FOSH) 3D printer, whose full specifications are released under the GNU General Public License, which is capable of replicating itself by printing many of its own (plastic) parts to create more machines. RepRaps have already been shown to be able to print circuit boards and metal parts. The most popular 3D printer in the world is the Prusa i3, a RepRap printer.

RepRap version 2.0 (Mendel)

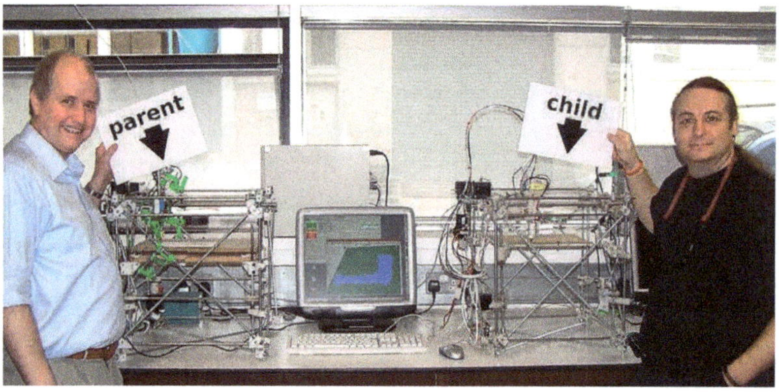

All of the plastic parts for the machine on the right were produced by the machine on the left. Adrian Bowyer (left) and Vik Olliver (right) are members of the RepRap project.

Because of the FOSH aims of RepRap, many related projects have used their design for inspiration, creating an ecosystem of related or derivative 3D printers, most of which are also open source designs. The availability of these open source designs means that variants of 3D printers are easy to invent. The quality and complexity of printer designs, however, as well as the quality of kit or finished products, varies greatly from project to project. This rapid development of open source 3D printers is gaining interest in many spheres as it enables hyper-customization and the use of public domain designs to fabricate open source appropriate technology. This technology can also assist initiatives in sustainable development since technologies are easily and economically made from resources available to local communities.

The cost of 3D printers has decreased dramatically since about 2010, with machines that used to cost $20,000 now costing less than $1,000. For instance, as of 2013, several companies and individuals are selling parts to build various RepRap designs, with prices starting at about €400 / US$500. The open source Fab@Home project has developed printers for general use with anything that can be squirted through a nozzle, from chocolate to silicone sealant and chemical reactants. Printers following the project's designs have been available from suppliers in kits or in pre-assembled form since 2012 at prices in the US$2000 range. Several new 3D printers are aimed at the small, inexpensive market including the mUVe3D and Lumifold. Rapide 3D has designed a professional grade crowdsourced 3D-printer costing $1499 which has no fumes nor constant rattle during use. The 3Doodler, 3D printing pen, raised $2.3 million on Kickstarter with the pens selling at $99, though the 3D Doodler has been criticized for being more of a crafting pen than a 3D printer.

As the costs of 3D printers have come down they are becoming more appealing financially to use for self-manufacturing of personal products. In addition, 3D printing products at home may reduce the environmental impacts of manufacturing by reducing material use and distribution impacts.

In addition, several RecycleBots such as the commercialized Filastruder have been designed and fabricated to convert waste plastic, such as shampoo containers and milk jugs, into inexpensive RepRap filament. There is some evidence that using this approach of distributed recycling is better for the environment.

The development and hyper-customization of the RepRap-based 3D printers has produced a new category of printers suitable for small business and consumer use. Manufacturers such as Solidoodle, Robo 3D, RepRapPro and Pirx 3D have introduced models and kits priced at less than $1,000, thousands less than they were in September 2012. Depending on the application, the print resolution and speed of manufacturing lies somewhere between a personal printer and an industrial printer. A list of printers with pricing and other information is maintained. Most recently delta robots, like the TripodMaker, have been utilized for 3D printing to increase fabrication speed further. For delta 3D printers, due to its geometry and differentiation movements, the accuracy of the print depends on the position of the printer head.

Some companies are also offering software for 3D printing, as a support for hardware manufactured by other companies.

Large 3D Printers

Large 3D printers have been developed for industrial, education, and demonstrative uses. A large delta-style 3D printer was built in 2014 by SeeMeCNC. The printer is capable of making an object with diameter of up to 4 feet (1.2 m) and up to 10 feet (3.0 m) in height. It also uses plastic pellets as the raw material instead of the typical plastic filaments used in other 3D printers.

The BigRep One.1 with its 1 m^3 volume.

Another type of large printer is Big Area Additive Manufacturing (BAAM). The goal is to develop printers that can produce a large object in high speed. A BAAM machine of Cincinnati Incorporated can produce an object at the speeds 200-500 times faster than typical 3D printers available in 2014. Another BAAM machine is being developed by Lockheed Martin with an aim to print long objects of up to 100 feet (30 m) to be used in aerospace industries.

Microscale and Nanoscale 3D Printing

Microelectronic device fabrication methods can be employed to perform the 3D printing of nanoscale-size objects. Such printed objects are typically grown on a solid substrate, e.g. silicon wafer, to which they adhere after printing as they are too small and fragile to be manipulated post-construction.

In one technique, 3D nanostructures can be printed by physically moving a dynamic stencil mask during the material deposition process, somewhat analogous to the extrusion method of traditional 3D printers. Programmable-height nanostructures with resolutions as small as 10 nm have been produced in this fashion, by metallic physical vapor deposition through a piezo-actuator controlled stencil mask having a milled nanopore in a silicon nitride membrane.

Another method enhances the photopolymerization process on a much smaller scale, using finely-focused lasers controlled by adjustable mirrors. This method has produced objects with feature resolutions of 100 nm.

Manufacturing Applications

Three-dimensional printing makes it as cheap to create single items as it is to produce thousands and thus undermines economies of scale. It may have as profound an impact on the world as the coming of the factory did....Just as nobody could have predicted the impact of the steam engine in 1750—or the printing press in 1450, or the transistor in 1950—it is impossible to foresee the long-term impact of 3D printing. But the technology is coming, and it is likely to disrupt every field it touches.

—The Economist, in a February 10, 2011 leader

AM technologies found applications starting in the 1980s in product development, data visualization, rapid prototyping, and specialized manufacturing. Their expansion into production (job production, mass production, and distributed manufacturing) has been under development in the decades since. Industrial production roles within the metalworking industries achieved significant scale for the first time in the early 2010s. Since the start of the 21st century there has been a large growth in the sales of AM machines, and their price has dropped substantially.According to Wohlers Associates, a consultancy, the market for 3D printers and services was worth $2.2 billion worldwide in 2012, up 29% from 2011. McKinsey predicts that additive manufacturing could have an economic impact of $550 billion annually by 2025. There are many applications for AM technologies, including architecture, construction (AEC), industrial design, automotive, aerospace, military, engineering, dental and medical industries, biotech (human tissue replacement), fashion, footwear, jewelry, eyewear, education, geographic information systems, food, and many other fields.

Additive manufacturing's earliest applications have been on the toolroom end of the manufacturing spectrum. For example, rapid prototyping was one of the earliest additive variants, and its mission was to reduce the lead time and cost of developing prototypes of new parts and devices, which was earlier only done with subtractive toolroom methods such as cnc milling and turning, and precision grinding, far more accurate than 3d printing with accuracy down to 0.00005 and creating better quality parts faster, but sometimes too expensive for low accuracy prototype parts. With technological advances in additive manufacturing, however, and the dissemination of those advances into the business world, additive methods are moving ever further into the production end of manufacturing in creative and sometimes unexpected ways. Parts that were formerly the sole province of subtractive methods can now in some cases be made more profitably via additive ones. In addition, new developments in RepRap technology allow the same device to perform both additive and subtractive manufacturing by swapping magnetic-mounted tool heads.

Cloud-Based Additive Manufacturing

Additive manufacturing in combination with cloud computing technologies allows decentralized and geographically independent distributed production. Cloud-based additive manufacturing refers to a service-oriented networked manufacturing model in which service consumers are able to build parts through Infrastructure-as-a-Service (IaaS), Platform-as-a-Service (PaaS), Hardware-as-a-Service (HaaS), and Software-as-a-Service (SaaS). Distributed manufacturing as such is carried out by some enterprises; there is also a services like 3D Hubs that put people needing 3D printing in contact with owners of printers. Some companies offer on-line 3D printing services to both commercial and private customers, working from 3D designs uploaded to the company website. 3D-printed designs are either shipped to the customer or picked up from the service provider.

Mass Customization

Miniature face models (from FaceGen) produced using Ceramic Based material on a Full Colour 3D Inkjet Printer

Companies have created services where consumers can customize objects using simplified web based customisation software, and order the resulting items as 3D printed unique objects. This now allows consumers to create custom cases for their mobile phones. Nokia has released the 3D designs for its case so that owners can customize their own case and have it 3D printed.

Rapid Manufacturing

Advances in RP technology have introduced materials that are appropriate for final manufacture, which has in turn introduced the possibility of directly manufacturing finished components. One advantage of 3D printing for rapid manufacturing lies in the relatively inexpensive production of small numbers of parts. Rapid manufacturing is a new method of manufacturing and many of its processes remain unproven. 3D printing is now entering the field of rapid manufacturing and was identified as a "next level" technology by many experts in a 2009 report. One of the most promising processes looks to be the adaptation of selective laser sintering (SLS), or direct metal laser sintering (DMLS) some of the better-established rapid prototyping methods. As of 2006, however, these techniques were still very much in their infancy, with many obstacles to be overcome before RM could be considered a realistic manufacturing method.

Rapid Prototyping

Industrial 3D printers have existed since the early 1980s and have been used extensively for rapid prototyping and research purposes. These are generally larger machines that use proprietary powdered metals, casting media (e.g. sand), plastics, paper or cartridges, and are used for rapid prototyping by universities and commercial companies.

Research

3D printing can be particularly useful in research labs due to its ability to make specialized, bespoke geometries. In 2012 a proof of principle project at the University of Glasgow, UK, showed that it is possible to use 3D printing techniques to assist in the production of chemical compounds. They first printed chemical reaction vessels, then used the printer to deposit reactants into them. They have produced new compounds to verify the validity of the process, but have not pursued anything with a particular application.

Food

Additive manufacturing of food is being developed by squeezing out food, layer by layer, into three-dimensional objects. A large variety of foods are appropriate candidates, such as chocolate and candy, and flat foods such as crackers, pasta, and pizza. NASA has considered the versatility of the concept, awarding a contract to the Systems and Materials Research Consultancy to study the feasibility of printing food in space. One of the problems with food printing is the nature of the texture of a food. For example, foods that are not strong enough to be filed are not appropriate for 3D printing.

Medical Applications

Professor Leroy Cronin of Glasgow University proposed in a 2012 TED Talk that it was possible to use chemical inks to print medicine. Similarly, 3D printing has been considered as a method of implanting stem cells capable of generating new tissues and organs in living humans. With their ability to transform into any other kind of cell in the human body, stem cells offer huge potential in 3D bio-printing.

A printing based on fused filament fabrication (FFF) approach has been already implemented for the creation of microstructures having an internal 3D microstructure geometry. These objects can be produced without any sacrificial structures or additional support materials, just by precisely tuning the nozzle heating, fan cooling and translation velocity parameters. The manufactured microporous structures out of polylactic acid (PLA) can have fully controllable porosity (20%–60%). Such scaffolds could serve as biomedical templates for cell culturing, as well as biodegradable implants for tissue engineering.

Agile Tooling

Agile tooling is a term used to describe the process of using modular means to design tooling that is produced by additive manufacturing or 3D printing methods to enable quick prototyping and responses to tooling and fixture needs. Agile tooling uses a cost effective and high quality method to quickly respond to customer and market needs. It can be used in hydro-forming, stamping, injection molding and other manufacturing processes.

Industrial Applications

Apparel

3D printing has entered the world of clothing with fashion designers experimenting with 3D-printed bikinis, shoes, and dresses. In commercial production Nike is using 3D printing to prototype and manufacture the 2012 Vapor Laser Talon football shoe for players of American football, and New Balance is 3D manufacturing custom-fit shoes for athletes.

3D printing has come to the point where companies are printing consumer grade eyewear with on-demand custom fit and styling (although they cannot print the lenses). On-demand customization of glasses is possible with rapid prototyping.

However, comment has been made in academic circles as to the potential limitation of the human acceptance of such mass customised apparel items due to the potential reduction of brand value communication.

In the world of high fashion courtiers such as Karl Lagerfeld designing for Chanel, Iris van Herpen and Noa Raviv working with technology from Stratasys, have employed and featured 3d printing in their collections. Selections from theie lines and other working with 3d printing were showcased at the 2016 Metropolitan Museum of Art Anna Wintour Costume Center, exhibition Manus X Machina.

Automotive Industry

In early 2014, the Swedish supercar manufacturer, Koenigsegg, announced the One:1, a supercar that utilizes many components that were 3D printed. In the limited run of vehicles Koenigsegg produces, the One:1 has side-mirror internals, air ducts, titanium exhaust components, and complete turbocharger assemblies that were 3D printed as part of the manufacturing process.

Urbee is the name of the first car in the world car mounted using the technology 3D printing (its bodywork and car windows were "printed"). Created in 2010 through the partnership between the US engineering group Kor Ecologic and the company Stratasys (manufacturer of printers Stratasys 3D), it is a hybrid vehicle with futuristic look.

The Audi RSQ was made with rapid prototyping industrial KUKA robots

In 2014, Local Motors debuted Strati, a functioning vehicle that was entirely 3D Printed using ABS plastic and carbon fiber, except the powertrain. In 2015, the company produced another iteration known as the LM3D Swim that was 80 percent 3D-printed. In 2016, the company has used 3D printing in the creation of automotive parts, such ones used in Olli, a self-driving vehicle developed by the company.

In May 2015 Airbus announced that its new Airbus A350 XWB included over 1000 components manufactured by 3D printing.

3D printing is also being utilized by air forces to print spare parts for planes. In 2015, a Royal Air Force Eurofighter Typhoon fighter jet flew with printed parts. The United States Air Force has begun to work with 3D printers, and the Israeli Air Force has also purchased a 3D printer to print spare parts.

Construction

The use of 3D printing to produce scale models within architecture and construction has steadily increased in popularity as the cost of 3D printers has reduced. This has enabled faster turn around of such scale models and allowed a steady increase in the speed of production and the complexity of the objects being produced.

Construction 3D printing, the application of 3D printing to fabricate construction components or entire buildings has been in development since the mid 1990s, development of new technologies has steadily gained pace since 2012 and the sub-sector of 3D printing is beginning to mature.

Firearms

In 2012, the US-based group Defense Distributed disclosed plans to "design a working plastic gun that could be downloaded and reproduced by anybody with a 3D printer." Defense Distributed has also designed a 3D printable AR-15 type rifle lower receiver (capable of lasting more than 650 rounds) and a 30-round M16 magazine. The AR-15 has multiple receivers (both an upper and lower receiver), but the legally controlled part is the one that is serialized (the lower, in the AR-15's case). Soon after Defense Distributed succeeded in designing the first working blueprint to produce a plastic gun with a 3D printer in May 2013, the United States Department of State demanded that they remove the instructions from their website. After Defense Distributed released their plans, questions were raised regarding the effects that 3D printing and widespread consumer-level CNC machining may have on gun control effectiveness.

In 2014, a man from Japan became the first person in the world to be imprisoned for making 3D printed firearms. Yoshitomo Imura posted videos and blueprints of the gun online and was sentenced to jail for two years. Police found at least two guns in his household that were capable of firing bullets.

Medical

3D printing has been used to print patient specific implant and device for medical use. Successful operations include a titanium pelvis implanted into a British patient, titanium lower jaw transplanted to a Belgian patient, and a plastic tracheal splint for an American infant. The hearing aid and dental industries are expected to be the biggest area of future development using the custom 3D printing technology. In March 2014, surgeons in Swansea used 3D printed parts to rebuild the face of a motorcyclist who had been seriously injured in a road accident. Research is also being conducted on methods to bio-print replacements for lost tissue due to arthritis and cancer.

3D printing technology can now be used to make exact replicas of organs. The printer uses images from patients' MRI or CT scan images as a template and lays down layers of rubber or plastic.

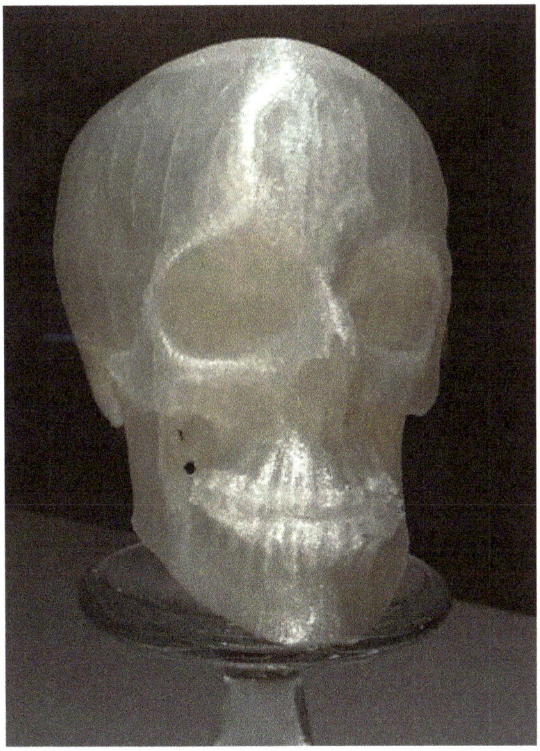

3D printed human skull from computed computer tomography data

Medical Devices

In October 24, 2014, a five-year-old girl born without fully formed fingers on her left hand became the first child in the UK to have a prosthetic hand made with 3D printing technology. Her hand was designed by US-based E-nable, an open source design organisation which uses a network of volunteers to design and make prosthetics mainly for children. The prosthetic hand was based on a plaster cast made by her parents. A boy named Alex was also born with a missing arm from just above the elbow. The team was able to use 3D printing to upload an e-NABLE Myoelectric arm that runs off of servos and batteries that are actuated by the electromyography muscle. With the use of 3D printers, E-NABLE has so far distributed more than 400 plastic hands to children.

Printed prosthetics have been used in rehabilitation of crippled animals. In 2013, a 3D printed foot let a crippled duckling walk again. In 2014 a chihuahua born without front legs was fitted with a harness and wheels created with a 3D printer. 3D printed hermit crab shells let hermit crabs inhabit a new style home. A prosthetic beak was another tool developed by the use of 3D printing to help aid a bald eagle named Beauty, whose beak was severely mutilated from a shot in the face. Since 2014, commercially available titanium knee implants made with 3D printer for dogs have been used to restore the animals› mobility. Over 10,000 dogs in Europe and United States have been treated after only one year.

In February 2015, FDA approved the marketing of a surgical bolt which facilitates less-invasive foot surgery and eliminates the need to drill through bone. The 3D printed titanium device, ‹FastForward Bone Tether Plate› is approved to use in correction surgery to treat bunion. In October 2015, the group of Professor Andreas Herrmann at the University of Groningen has developed the first 3D printable resins with antimicrobial properties. Employing stereolithography, quaternary ammonium groups are incorporated into dental appliances that kill bacteria on contact. This type of material can be further applied in medical devices and implants.

In 2013, researchers at the University of Hasselt, in Belgium had successfully printed a new jawbone for an 83-year-old Belgian woman.

3D printing has been used to produce prosthetic beaks for eagles, a Brazilian goose named Victoria, and a Costa Rican toucan called Grecia.

Bio-Printing

As of 2012, 3D bio-printing technology has been studied by biotechnology firms and academia for possible use in tissue engineering applications in which organs and body parts are built using inkjet techniques. In this process, layers of living cells are deposited onto a gel medium or sugar matrix and slowly built up to form three-dimensional structures including vascular systems. The first production system for 3D tissue printing was delivered in 2009, based on NovoGen bioprinting technology. Several terms have been used to refer to this field of research: organ printing, bio-printing, body part printing, and computer-aided tissue engineering, among others. The possibility of using 3D tissue printing to create soft tissue architectures for reconstructive surgery is also being explored.

In 2013, Chinese scientists began printing ears, livers and kidneys, with living tissue. Researchers in China have been able to successfully print human organs using specialized 3D bio printers that use living cells instead of plastic. Researchers at Hangzhou Dianzi University designed the 3D bio printer dubbed the Regenovo. Xu Mingen, Regenovo›s developer, said that it takes the printer under an hour to produce either a mini liver sample or a four to five inch ear cartilage sample. Xu also predicted that fully functional printed organs may be possible within the next ten to twenty years.

Pills

The first pill manufactured by 3D printing was approved by the FDA in August 2015. Binder-jetting into a powder bed of the drug allows very porous pills to be produced, which enables high drug doses in a single pill which dissolves quickly and can be ingested easily. This has been demonstrated for Spritam, a reformulation of levetiracetam for the treatment of epilepsy.

Computers and Robots

3D printing can also be used to make laptops and other computers and cases. For example, Novena and VIA OpenBook standard laptop cases. I.e. a Novena motherboard can be bought and be used in a printed VIA OpenBook case.

Open-source robots are built using 3D printers. Double Robotics grant access to their technology (an open SDK). On the other hand, 3&DBot is an Arduino 3D printer-robot with wheels and ODOI is a 3D printed humanoid robot.

Space

The Zero-G Printer, the first 3D printer designed to operate in zero gravity, was built under a joint partnership between NASA Marshall Space Flight Center (MSFC) and Made In Space, Inc. In September 2014, SpaceX delivered the zero-gravity 3D printer to the International Space Station (ISS). On December 19, 2014, NASA emailed CAD drawings for a socket wrench to astronauts aboard the ISS, who then printed the tool using its 3D printer. Applications for space offer the ability to print parts or tools on-site, as opposed to using rockets to bring along pre-manufactured items for space missions to human colonies on the moon, Mars, or elsewhere. The European Space Agency plans to deliver its new Portable On-Board 3D Printer (POP3D for short) to the International Space Station by June 2015, making it the second 3D printer in space.

Furthermore, the Sinterhab project is researching a lunar base constructed by 3D printing using lunar regolith as a base material. Instead of adding a binding agent to the regolith, researchers are experimenting with microwave sintering to create solid blocks from the raw material.

Similar researches and projects like these could allow faster construction for lower costs, and has been investigated for construction of off-Earth habitats.

Sociocultural Applications

An example of 3D printed limited edition jewellery. This necklace is made of glassfiber-filled dyed nylon. It has rotating linkages that were produced in the same manufacturing step as the other parts

In 2005, a rapidly expanding hobbyist and home-use market was established with the inauguration of the open-source RepRap and Fab@Home projects. Virtually all home-use 3D printers released to-date have their technical roots in the ongoing RepRap Project and associated open-source software initiatives. In distributed manufacturing, one study has found that 3D printing could become a mass market product enabling consumers to save money associated with purchasing common household objects. For example, instead of going to a store to buy an object made in a factory by injection molding (such as a measuring cup or a funnel), a person might instead print it at home from a downloaded 3D model.

Art

In 2005, academic journals had begun to report on the possible artistic applications of 3D printing technology, being used by artists such as Martin John Callanan at The Bartlett school of architecture. By 2007 the mass media followed with an article in the Wall Street Journal and Time Magazine, listing a 3D printed design among their 100 most influential designs of the year. During the 2011 London Design Festival, an installation, curated by Murray Moss and focused on 3D Printing, was held in the Victoria and Albert Museum (the V&A). The installation was called *Industrial Revolution 2.0: How the Material World will Newly Materialize*.

Some of the recent developments in 3D printing were revealed at the 3DPrintshow in London, which took place in November 2013 and 2014. The art section had in exposition artworks made with 3D printed plastic and metal. Several artists such as Joshua Harker, Davide Prete, Sophie Kahn, Helena Lukasova, Foteini Setaki showed how 3D printing can modify aesthetic and art processes. One part of the show focused on ways in which 3D printing can advance the medical field. The underlying theme of these advances was that these printers can be used to create parts that are printed with specifications to meet each individual. This makes the process safer and more efficient. One of these advances is the use of 3D printers to produce casts that are created to mimic the bones that they are supporting. These custom-fitted casts are open, which allow the wearer to scratch any itches and also wash the damaged area. Being open also allows for open ventilation. One of the best features is that they can be recycled to create more casts.

3D printing is becoming more popular in the customisable gifts industry, with products such as personalized mobile phone cases and dolls, as well as 3D printed chocolate.

The use of 3D scanning technologies allows the replication of real objects without the use of moulding techniques that in many cases can be more expensive, more difficult, or too invasive to be performed, particularly for precious or delicate cultural heritage artifacts where direct contact with the moulding substances could harm the original object›s surface.

Critical making refers to the hands on productive activities that link digital technologies to society. It is invented to bridge the gap between creative physical and conceptual exploration. The term was popularized by Matt Ratto, an Assistant Professor and director of the Critical Making lab in the Faculty of Information at the University of Toronto. Ratto describes one of the main goals of critical as to use material forms of engagement with technologies to supplement and extend critical reflection and, in doing so, to reconnect our lived experiences with technologies to social and conceptual critique. The main focus of critical making is open design, which includes, in addition to 3D printing technologies, also other digital software and hardware. People usually reference spectacular design when explaining critical making.

Communication

Employing additive layer technology offered by 3D printing, Terahertz devices which act as waveguides, couplers and bends have been created. The complex shape of these devices could not be achieved using conventional fabrication techniques. Commercially available professional grade printer EDEN 260V was used to create structures with minimum feature size of 100 μm. The printed structures were later DC sputter coated with gold (or any other metal) to create a Terahertz Plasmonic Device. In 2016 artist/scientist Janine Carr Created the first 3d printed vocal percussion (beatbox) as a waveform, with the ability to play the soundwave by laser, along with four vocalised emotions these were also playable by laser.

Domestic Use

Some early consumer examples of 3d printing include the 64DD released in 1999 in Japan. As of 2012, domestic 3D printing was mainly practiced by hobbyists and enthusiasts. However, little was used for practical household applications, for example, ornamental objects. Some practical examples include a working clock and gears printed for home woodworking machines among other purposes. Web sites associated with home 3D printing tended to include backscratchers, coat hooks, door knobs, etc.

The open source Fab@Home project has developed printers for general use. They have been used in research environments to produce chemical compounds with 3D printing technology, including new ones, initially without immediate application as proof of principle. The printer can print with anything that can be dispensed from a syringe as liquid or paste. The developers of the chemical application envisage both industrial and domestic use for this technology, including enabling users in remote locations to be able to produce their own medicine or household chemicals.

3D printing is now working its way into households, and more and more children are being introduced to the concept of 3D printing at earlier ages. The prospects of 3D printing are growing, and as more people have access to this new innovation, new uses in households will emerge.

The OpenReflex SLR film camera was developed for 3D printing as an open-source student project.

Education and Research

High School students from Wyomissing Area Jr/Sr High School in Pennsylvania, United States present their use of 3D Printing in the classroom

3D printing, and open source 3D printers in particular, are the latest technology making inroads into the classroom. 3D printing allows students to create prototypes of items without the use of expensive tooling required in subtractive methods. Students design and produce actual models they can hold. The classroom environment allows students to learn and employ new applications for 3D printing.RepRaps, for example, have already been used for an educational mobile robotics platform.

Some authors have claimed that 3D printers offer an unprecedented "revolution" in STEM education.The evidence for such claims comes from both the low cost ability for rapid prototyping in the classroom by students, but also the fabrication of low-cost high-quality scientific equipment from open hardware designs forming open-source labs. Engineering and design principles are explored as well as architectural planning. Students recreate duplicates of museum items such as fossils and historical artifacts for study in the classroom without possibly damaging sensitive collections. Other students interested in graphic designing can construct models with complex working parts easily. 3D printing gives students a new perspective with topographic maps. Science students can study cross-sections of internal organs of the human body and other biological specimens. And chemistry students can explore 3D models of molecules and the relationship within chemical compounds.

According to a recent paper by Kostakis et al., 3D printing and design can electrify various literacies and creative capacities of children in accordance with the spirit of the interconnected, information-based world. Future applications for 3D printing might include creating open-source scientific equipment.

Environmental Use

In Bahrain, large-scale 3D printing using a sandstone-like material has been used to create unique coral-shaped structures, which encourage coral polyps to colonize and regenerate damaged reefs. These structures have a much more natural shape than other structures used to create artificial reefs, and, unlike concrete, are neither acid nor alkaline with neutral pH.

Cultural Heritage

In the last several years 3D printing has been intensively used by in the cultural heritage field for preservation, restoration and dissemination purposes. Many Europeans and North American Museums have purchased 3D printers and actively recreate missing pieces of their relics.

The Metropolitan Museum of Art and the British Museum have started using their 3D printers to create museum souvenirs that are available in the museum shops. Other museums, like the National Museum of Military History and Varna Historical Museum, have gone further and sell through the online platform Threeding digital models of their artifacts, created using Artec 3D scanners, in 3D printing friendly file format, which everyone can 3D print at home.

3D printed sculpture of the Egyptian Pharaoh Merankhre Mentuhotep shown at Threeding

Specialty Materials

Consumer grade 3D printing has resulted in new materials that have been developed specifically for 3D printers. For example, filament materials have been developed to imitate wood in its appearance as well as its texture. Furthermore, new technologies, such as infusing carbon fiber into printable plastics, allowing for a stronger, lighter material. In addition to new structural materials that have been developed due to 3D printing, new technologies have allowed for patterns to be applied directly to 3D printed parts. Iron oxide-free Portland cement powder has been used to create architectural structures up to 9 feet in height.

Legal Aspects

Intellectual Property

3D printing has existed for decades within certain manufacturing industries where many legal regimes, including patents, industrial design rights, copyright, and trademark may apply. However, there is not much jurisprudence to say how these laws will apply if 3D printers become mainstream and individuals and hobbyist communities begin manufacturing items for personal use, for non-profit distribution, or for sale.

Any of the mentioned legal regimes may prohibit the distribution of the designs used in 3D printing, or the distribution or sale of the printed item. To be allowed to do these things, where an active intellectual property was involved, a person would have to contact the owner and ask for a licence, which may come with conditions and a price. However, many patent, design and copyright laws contain a standard limitation or exception for 'private', 'non-commercial' use of inventions, designs or works of art protected under intellectual property (IP). That standard limitation or exception may leave such private, non-commercial uses outside the scope of IP rights.

Patents cover inventions including processes, machines, manufactures, and compositions of matter and have a finite duration which varies between countries, but generally 20 years from the date of application. Therefore, if a type of wheel is patented, printing, using, or selling such a wheel could be an infringement of the patent.

Copyright covers an expression in a tangible, fixed medium and often lasts for the life of the author plus 70 years thereafter. If someone makes a statue, they may have copyright on the look of that statue, so if someone sees that statue, they cannot then distribute designs to print an identical or similar statue.

When a feature has both artistic (copyrightable) and functional (patentable) merits, when the question has appeared in US court, the courts have often held the feature is not copyrightable unless it can be separated from the functional aspects of the item. In other countries the law and the courts may apply a different approach allowing, for example, the design of a useful device to be registered (as a whole) as an industrial design on the understanding that, in case of unauthorized copying, only the non-functional features may be claimed under design law whereas any technical features could only be claimed if covered by a valid patent.

Gun Legislation and Administration

The US Department of Homeland Security and the Joint Regional Intelligence Center released a memo stating that "significant advances in three-dimensional (3D) printing capabilities, availability of free digital 3D printable files for firearms components, and difficulty regulating file sharing may present public safety risks from unqualified gun seekers who obtain or manufacture 3D printed guns," and that "proposed legislation to ban 3D printing of weapons may deter, but cannot completely prevent their production. Even if the practice is prohibited by new legislation, online distribution of these 3D printable files will be as difficult to control as any other illegally traded music, movie or software files."

Internationally, where gun controls are generally stricter than in the United States, some commentators have said the impact may be more strongly felt, as alternative firearms are not as easily obtainable. Officials in the United Kingdom have noted that producing a 3D printed gun would be illegal under their gun control laws. Europol stated that criminals have access to other sources of weapons, but noted that as the technology improved the risks of an effect would increase. Downloads of the plans from the UK, Germany, Spain, and Brazil were heavy.

Attempting to restrict the distribution over the Internet of gun plans has been likened to the futility of preventing the widespread distribution of DeCSS which enabled DVD ripping. After the US government had Defense Distributed take down the plans, they were still widely available via The Pirate Bay and other file sharing sites. Some US legislators have proposed regulations on 3D printers, to prevent them being used for printing guns. 3D printing advocates have suggested that such regulations would be futile, could cripple the 3D printing industry, and could infringe on free speech rights, with early pioneer of 3D printing Professor Hod Lipson suggesting that gunpowder could be controlled instead.

Safety

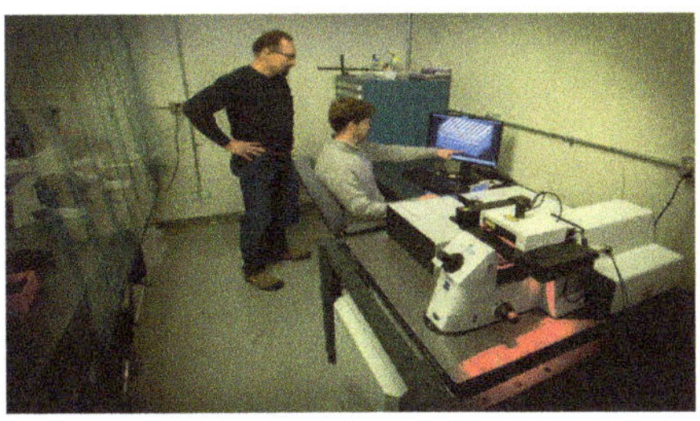

A video on research done on printer emissions

3D printers have been noted to be an environmental hazard due to them emitting microscopic particles and chemicals that have been linked to asthma. A National Institute for Occupational Safety and Health (NIOSH) report notes these emissions peaked a few minutes after printing started and returned to baseline levels 100 minutes after printing ended. The problem was reduced by using manufacturer-supplied covers and full enclosures, using proper ventilation, keeping workers away from the printer while wearing respirators, turning off the printer if it jammed, and using lower emission printers and filaments.

Impact

Additive manufacturing, starting with today's infancy period, requires manufacturing firms to be flexible, ever-improving users of all available technologies to remain competitive. Advocates of additive manufacturing also predict that this arc of technological development will counter globalization, as end users will do much of their own manufacturing rather than engage in trade to buy products from other people and corporations. The real integration of the newer additive technologies into commercial production, however, is more a matter of complementing traditional subtractive methods rather than displacing them entirely.

Social Change

Since the 1950s, a number of writers and social commentators have speculated in some depth about the social and cultural changes that might result from the advent of commercially affordable additive manufacturing technology. Amongst the more notable ideas to have emerged from these inquiries has been the suggestion that, as more and more 3D printers start to enter people's homes, the conventional relationship between the home and the workplace might get further eroded. Likewise, it has also been suggested that, as it becomes easier for businesses to transmit designs for new objects around the globe, so the need for high-speed freight services might also become less. Finally, given the ease with which certain objects can now be replicated, it remains to be seen whether changes will be made to current copyright legislation so as to protect intellectual property rights with the new technology widely available.

As 3D printers became more accessible to consumers, online social platforms have developed to support the community. This includes websites that allow users to access information such as how to build a 3D printer, as well as social forums that discuss how to improve 3D print quality and discuss 3D printing news, as well as social media websites that are dedicated to share 3D models. RepRap is a wiki based website that was created to hold all information on 3d printing, and has developed into a community that aims to bring 3D printing to everyone. Furthermore, there are other sites such as Pinshape, Thingiverse and MyMiniFactory, which were created initially to allow users to post 3D files for anyone to print, allowing for decreased transaction cost of sharing 3D files. These websites have allowed greater social interaction between users, creating communities dedicated to 3D printing.

Some call attention to the conjunction of Commons-based peer production with 3D printing and other low-cost manufacturing techniques. The self-reinforced fantasy of a system of eternal growth can be overcome with the development of economies of scope, and here, society can play an important role contributing to the raising of the whole productive structure to a higher plateau of more sustainable and customized productivity. Further, it is true that many issues, problems, and threats arise due to the democratization of the means of production, and especially regarding the physical ones. For instance, the recyclability of advanced nanomaterials is still questioned; weapons manufacturing could become easier; not to mention the implications for counterfeiting and on IP. It might be maintained that in contrast to the industrial paradigm whose competitive dynamics were about economies of scale, Commons-based peer production 3D printing could develop economies of scope. While the advantages of scale rest on cheap global transportation, the economies of scope share infrastructure costs (intangible and tangible productive resources), taking advantage of the capabilities of the fabrication tools. And following Neil Gershenfeldin that some of the least developed parts of the world need some of the most advanced technologies, Commons-based peer production and 3D printing may offer the necessary tools for thinking globally but acting locally in response to certain needs.

Larry Summers wrote about the devastating consequences of 3D printing and other technologies (robots, artificial intelligence, etc.) for those who perform routine tasks. In his view, already there are more American men on disability insurance than doing production work in manufacturing. And the trends are all in the wrong direction, particularly for the less skilled, as the capacity of capital embodying artificial intelligence to replace white-collar as well as blue-collar work will increase rapidly in the years ahead. Summers recommends more vigorous cooperative efforts to address the myriad devices (e.g., tax havens, bank secrecy, money laundering, and regulatory arbitrage) enabling the holders of great wealth to avoid paying income and estate taxes, and to make it more difficult to accumulate great fortunes without requiring great social contributions in return, including: more vigorous enforcement of anti-monopoly laws, reductions in excessive protection for intellectual property, greater encouragement of profit-sharing schemes that may benefit workers and give them a stake in wealth accumulation, strengthening of collective bargaining arrangements, improvements in corporate governance, strengthening of financial regulation to eliminate subsidies to financial activity, easing of land-use restrictions that may cause the real estate of the rich to keep rising in value, better training for young people and retraining for displaced workers, and increased public and private investment in infrastructure development—e.g., in energy production and transportation.

Michael Spence wrote that Now comes a ... powerful, wave of digital technology that is replacing labor in increasingly complex tasks. This process of labor substitution and disintermediation has been underway for some time in service sectors—think of ATMs, online banking, enterprise resource planning, customer relationship management, mobile payment systems, and much more. This revolution is spreading to the production of goods, where robots and 3D printing are displacing labor. In his view, the vast majority of the cost of digital technologies comes at the start, in the design of hardware (e.g. 3D printers) and, more important, in creating the software that enables machines to carry out various tasks. Once this is achieved, the marginal cost of the hardware is relatively low (and declines as scale rises), and the marginal cost of replicating the software is essentially zero. With a huge potential global market to amortize the upfront fixed costs of design and testing, the incentives to invest in digital technologies are compelling. Spence believes that, unlike prior digital technologies, which drove firms to deploy underutilized pools of valuable labor around the world, the motivating force in the current wave of digital technologies is cost reduction via the replacement of labor. For example, as the cost of 3D printing technology declines, it is easy to imagine that production may become extremely local and customized. Moreover, production may occur in response to actual demand, not anticipated or forecast demand. Spence believes that labor, no matter how inexpensive, will become a less important asset for growth and employment expansion, with labor-intensive, process-oriented manufacturing becoming less effective, and that re-localization will appear in both developed and developing countries. In his view, production will not disappear, but it will be less labor-intensive, and all countries will eventually need to rebuild their growth models around digital technologies and the human capital supporting their deployment and expansion. Spence writes that the world we are entering is one in which the most powerful global flows will be ideas and digital capital, not goods, services, and traditional capital. Adapting to this will require shifts in mindsets, policies, investments (especially in human capital), and quite possibly models of employment and distribution.

Forbes investment pundits have predicted that 3D printing may lead to a resurgence of American Manufacturing, citing the small, creative companies that comprise the current industry landscape, and the lack of the necessary complex infrastructure in typical outsource markets.

3D printing is the process by which a three-dimensional structure is built through the various processes of additive manufacturing. Nanoscale 3D printing involves many of the same process, incorporated at a much smaller scale. To print a structure in the 5-400 μm scale, the precision of the 3D printing machine is improved greatly.

3D Printing and Laser Etching

A method pioneered in Seoul, South Korea uses a two-step process of 3D printing, using a 3D printing and laser etched plates. To be more precise at a nanoscale, the 3D printing process uses a laser etching machine, which etches into each plate the details needed for the segment of nanorobot. The plate is then transferred to the 3D printer, which fills the etched regions with the desired nanoparticle. The 3D printing process is repeated until the nanorobot is built from the bottom up. This 3D printing process has many benefits. First, it increases the overall accuracy of the printing process. Second, it has the potential to create functional segments of a nanorobot.

Two-photon lithography

The 3D printer uses a liquid resin, which is hardened at precisely the correct spots by a focused laser beam. The focal point of the laser beam is guided through the resin by movable mirrors and leaves behind a hardened line of solid polymer, just a few hundred nanometers wide. This fine resolution enables the creation of intricately structured sculptures as tiny as a grain of sand. This process takes place by using photoactive resins, which are hardened by the laser at an extremely small scale to create the structure. This process is quick by nanoscale 3D printing standards. Ultra-small features can be made with the 3D micro-fabrication technique used in multiphoton photopolymerisation. This approach uses a focused laser to trace the desired 3D object into a block of gel. Due to the nonlinear nature of photo excitation, the gel is cured to a solid only in the places where the laser was focused while the remaining gel is then washed away. Feature sizes of under 100 nm are easily produced, as well as complex structures with moving and interlocked parts.

NANOMEDICINE

Nanomedicine is the medical application of nanotechnology. Nanomedicine ranges from the medical applications of nanomaterials and biological devices, to nanoelectronic biosensors, and even possible future applications of molecular nanotechnology such as biological machines. Current problems for nanomedicine involve understanding the issues related to toxicity and environmental impact of nanoscale materials (materials whose structure is on the scale of nanometers, i.e. billionths of a meter).

Functionalities can be added to nanomaterials by interfacing them with biological molecules or structures. The size of nanomaterials is similar to that of most biological molecules and structures; therefore, nanomaterials can be useful for both in vivo and in vitro biomedical research and applications. Thus far, the integration of nanomaterials with biology has led to the development of diagnostic devices, contrast agents, analytical tools, physical therapy applications, and drug delivery vehicles.

Nanomedicine seeks to deliver a valuable set of research tools and clinically useful devices in the near future. The National Nanotechnology Initiative expects new commercial applications in the pharmaceutical industry that may include advanced drug delivery systems, new therapies, and in vivo imaging. Nanomedicine research is receiving funding from the US National Institutes of Health, including the funding in 2005 of a five-year plan to set up four nanomedicine centers. Nanomedicine sales reached \$16 billion in 2015, with a minimum of \$3.8 billion in nanotechnology R&D being invested every year. Global funding for emerging nanotechnology increased by 45% per year in recent years, with product sales exceeding \$1 trillion in 2013. As the nanomedicine industry continues to grow, it is expected to have a significant impact on the economy.

Drug Delivery

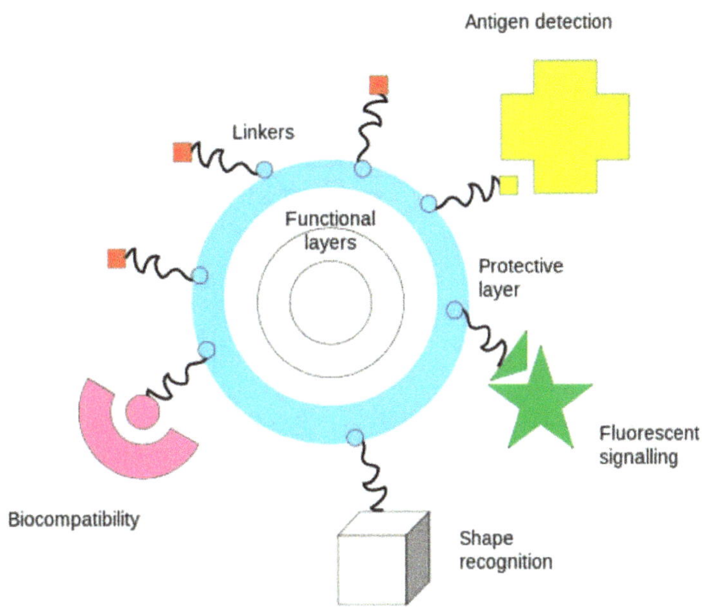

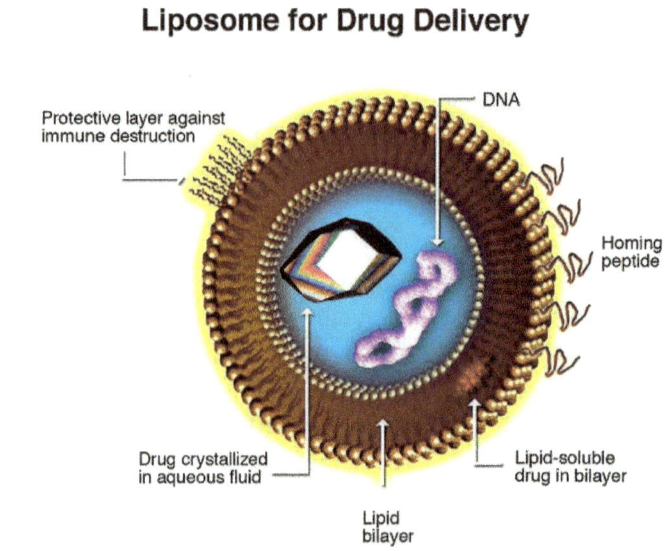

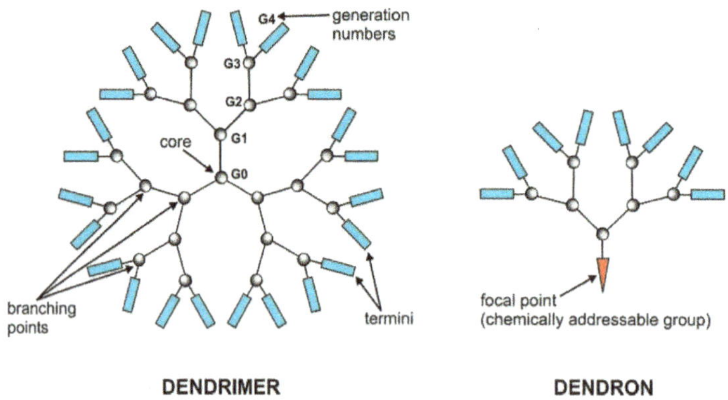

Nanoparticles *(top)*, liposomes*(middle)*, and dendrimers *(bottom)* are some nanomaterials being investigated for use in nanomedicine.

Nanotechnology has provided the possibility of delivering drugs to specific cells using nanoparticles.

The overall drug consumption and side-effects may be lowered significantly by depositing the active agent in the morbid region only and in no higher dose than needed. Targeted drug delivery is intended to reduce the side effects of drugs with concomitant decreases in consumption and treatment expenses. Drug delivery focuses on maximizing bioavailability both at specific places in the body and over a period of time. This can potentially be achieved by molecular targeting by nanoengineered devices. More than $65 billion are wasted each year due to poor bioavailability. A benefit of using nanoscale for medical technologies is that smaller devices are less invasive and can possibly be implanted inside the body, plus biochemical reaction times are much shorter. These devices are faster and more sensitive than typical drug delivery. The efficacy of drug delivery through nanomedicine is largely based upon: a) efficient encapsulation of the drugs, b) successful delivery of drug to the targeted region of the body, and c) successful release of the drug.

Drug delivery systems, lipid- or polymer-based nanoparticles, can be designed to improve the pharmacokinetics and biodistribution of the drug. However, the pharmacokinetics and pharmacodynamics of nanomedicine is highly variable among different patients. When designed to avoid the body's defence mechanisms, nanoparticles have beneficial properties that can be used to improve drug delivery. Complex drug delivery mechanisms are being developed, including the ability to get drugs through cell membranes and into cell cytoplasm. Triggered response is one way for drug molecules to be used more efficiently. Drugs are placed in the body and only activate on encountering a particular signal. For example, a drug with poor solubility will be replaced by a drug delivery system where both hydrophilic and hydrophobic environments exist, improving the solubility. Drug delivery systems may also be able to prevent tissue damage through regulated drug release; reduce drug clearance rates; or lower the volume of distribution and reduce the effect on non-target tissue. However, the biodistribution of these nanoparticles is still imperfect due to the complex host›s reactions to nano- and microsized materials and the difficulty in targeting specific organs in the body. Nevertheless, a lot of work is still ongoing to optimize and better understand the potential and limitations of nanoparticulate systems. While advancement of research proves that targeting and distribution can be augmented by nanoparticles, the dangers of nanotoxicity become an important next step in further understanding of their medical uses.

Nanoparticles can be used in combination therapy for decreasing antibiotic resistance or for their antimicrobial properties. Nanoparticles might also used to circumvent multidrug resistance (MDR) mechanisms.

Types of Systems Used

Two forms of nanomedicine that have already been tested in mice and are awaiting human testing will use gold nanoshells to help diagnose and treat cancer, along with liposomes as vaccine adjuvants and drug transport vehicles. Similarly, drug detoxification is also another application for nanomedicine which has shown promising results in rats. Advances in Lipid nanotechnology was also instrumental in engineering medical nanodevices and novel drug delivery systems as well as in developing sensing applications. Another example can be found in dendrimers and nanoporous materials. Another example is to use block co-polymers, which form micelles for drug encapsulation.

Polymeric nano-particles are a competing technology to lipidic (based mainly on Phospholipids) nano-particles. There is an additional risk of toxicity associated with polymers not widely studied or understood. The major advantages of polymers is stability, lower cost and predictable characterisation. However, in the patient›s body this very stability (slow degradation) is a negative factor. Phospholipids on the other hand are membrane lipids (already present in the body and surrounding each cell), have a GRAS (Generally Recognised As Safe) status from FDA and are derived from natural sources without any complex chemistry involved. They are not metabolised but rather absorbed by the body and the degradation products are themselves nutrients (fats or micronutrients).

Protein and peptides exert multiple biological actions in the human body and they have been identified as showing great promise for treatment of various diseases and disorders. These macromolecules are called biopharmaceuticals. Targeted and/or controlled delivery of these biopharmaceuticals using nanomaterials like nanoparticles and Dendrimers is an emerging field called nanobiopharmaceutics, and these products are called nanobiopharmaceuticals.

Another highly efficient system for microRNA delivery for example are nanoparticles formed by the self-assembly of two different microRNAs deregulated in cancer.

Another vision is based on small electromechanical systems; nanoelectromechanical systems are being investigated for the active release of drugs. Some potentially important applications include cancer treatment with iron nanoparticles or gold shells.Nanotechnology is also opening up new opportunities in implantable delivery systems, which are often preferable to the use of injectable drugs, because the latter frequently display first-order kinetics (the blood concentration goes up rapidly, but drops exponentially over time). This rapid rise may cause difficulties with toxicity, and drug efficacy can diminish as the drug concentration falls below the targeted range.

Applications

Some nanotechnology-based drugs that are commercially available or in human clinical trials include:

- Abraxane, approved by the U.S. Food and Drug Administration (FDA) to treat breast cancer, non-small- cell lung cancer (NSCLC) and pancreatic cancer, is the nanoparticle albumin bound paclitaxel.
- Doxil was originally approved by the FDA for the use on HIV-related Kaposi's sarcoma. It is now being used to also treat ovarian cancer and multiple myeloma. The drug is encased in liposomes, which helps to extend the life of the drug that is being distributed. Liposomes are self-assembling, spherical, closed colloidal structures that are composed of lipid bilayers that surround an aqueous space. The liposomes also help to increase the functionality and it helps to decrease the damage that the drug does to the heart muscles specifically.
- Onivyde, liposome encapsulated irinotecan to treat metastatic pancreatic cancer, was approved by FDA in October 2015.
- C-dots (Cornell dots) are the smallest silica-based nanoparticles with the size <10 nm. The particles are infused with organic dye which will light up with fluorescence. Clinical trial is underway since 2011 to use the C-dots as diagnostic tool to assist surgeons to identify the location of tumor cells.
- An early phase clinical trial using the platform of 'Minicell' nanoparticle for drug delivery have been tested on patients with advanced and untreatable cancer. Built from the membranes of mutant bacteria, the minicells were loaded with paclitaxel and coated with cetuximab, antibodies that bind the epidermal growth factor receptor (EGFR) which is often overexpressed in a number of cancers, as a ‹homing› device to the tumor cells. The tumor cells recognize the bacteria from which the minicells have been derived, regard it as invading

microorganism and engulf it. Once inside, the payload of anti-cancer drug kills the tumor cells. Measured at 400 nanometers, the minicell is bigger than synthetic particles developed for drug delivery. The researchers indicated that this larger size gives the minicells a better profile in side-effects because the minicells will preferentially leak out of the porous blood vessels around the tumor cells and do not reach the liver, digestive system and skin. This Phase 1 clinical trial demonstrated that this treatment is well tolerated by the patients. As a platform technology, the minicell drug delivery system can be used to treat a number of different cancers with different anti-cancer drugs with the benefit of lower dose and less side-effects.

- In 2014, a Phase 3 clinical trial for treating inflammation and pain after cataract surgery, and a Phase 2 trial for treating dry eye disease were initiated using nanoparticle loteprednol etabonate. In 2015, the product, KPI-121 was found to produce statistically significant positive results for the post-surgery treatment.

Cancer

Existing and potential drug nanocarriers have been reviewed.

Nanoparticles have high surface area to volume ratio. This allows for many functional groups to be attached to a nanoparticle, which can seek out and bind to certain tumor cells. Additionally, the small size of nanoparticles (10 to 100 nanometers), allows them to preferentially accumulate at tumor sites (because tumors lack an effective lymphatic drainage system). Limitations to conventional cancer chemotherapy include drug resistance, lack of selectivity, and lack of solubility. Nanoparticles have the potential to overcome these problems.

In photodynamic therapy, a particle is placed within the body and is illuminated with light from the outside. The light gets absorbed by the particle and if the particle is metal, energy from the light will heat the particle and surrounding tissue. Light may also be used to produce high energy oxygen molecules which will chemically react with and destroy most organic molecules that are next to them (like tumors). This therapy is appealing for many reasons. It does not leave a "toxic trail" of reactive molecules throughout the body (chemotherapy) because it is directed where only the light is shined and the particles exist. Photodynamic therapy has potential for a noninvasive procedure for dealing with diseases, growth and tumors. Kanzius RF therapy is one example of such therapy (nanoparticle hyperthermia) . Also, gold nanoparticles have the potential to join numerous therapeutic functions into a single platform, by targeting specific tumor cells, tissues and organs.

Visualization

In vivo imaging is another area where tools and devices are being developed. Using nanoparticle contrast agents, images such as ultrasound and MRI have a favorable distribution and improved contrast. This might be accomplished by self assembled biocompatible nanodevices that will detect, evaluate, treat and report to the clinical doctor automatically.

The small size of nanoparticles endows them with properties that can be very useful in oncology, particularly in imaging. Quantum dots (nanoparticles with quantum confinement properties, such as size-tunable light emission), when used in conjunction with MRI (magnetic resonance imaging), can produce exceptional images of tumor sites. Nanoparticles of cadmium selenide (quantum dots) glow when exposed to ultraviolet light. When injected, they seep into cancer tumors. The surgeon can see the glowing tumor, and use it as a guide for more accurate tumor removal.These nanoparticles are much brighter than organic dyes and only need one light source for excitation. This means that the use of fluorescent quantum dots could produce a higher contrast image and at a lower cost than today›s organic dyes used as contrast media. The downside, however, is that quantum dots are usually made of quite toxic elements.

Tracking movement can help determine how well drugs are being distributed or how substances are metabolized. It is difficult to track a small group of cells throughout the body, so scientists used to dye the cells. These dyes needed to be excited by light of a certain wavelength in order for them to light up. While different color dyes absorb different frequencies of light, there was a need for as many light sources as cells. A way around this problem is with luminescent tags. These tags are quantum dots attached to proteins that penetrate cell membranes. The dots can be random in size, can be made of bio-inert material, and they demonstrate the nanoscale property that color is size-dependent. As a result, sizes are selected so that the frequency of light used to make a group of quantum dots fluoresce is an even multiple of the frequency required to make another group incandesce. Then both groups can be lit with a single light source. They have also found a way to insert nanoparticles into the affected parts of the body so that those parts of the body will glow showing the tumor growth or shrinkage or also organ trouble.

Sensing

Nanotechnology-on-a-chip is one more dimension of lab-on-a-chip technology. Magnetic nanoparticles, bound to a suitable antibody, are used to label specific molecules, structures or microorganisms. Gold nanoparticles tagged with short segments of DNA can be used for detection of genetic sequence in a sample. Multicolor optical coding for biological assays has been achieved by embedding different-sized quantum dots into polymeric microbeads. Nanopore technology for analysis of nucleic acids converts strings of nucleotides directly into electronic signatures.

Sensor test chips containing thousands of nanowires, able to detect proteins and other biomarkers left behind by cancer cells, could enable the detection and diagnosis of cancer in the early stages from a few drops of a patient's blood. Nanotechnology is helping to advance the use of arthroscopes, which are pencil-sized devices that are used in surgeries with lights and cameras so surgeons can do the surgeries with smaller incisions. The smaller the incisions the faster the healing time which is better for the patients. It is also helping to find a way to make an arthroscope smaller than a strand of hair.

Research on nanoelectronics-based cancer diagnostics could lead to tests that can be done in pharmacies. The results promise to be highly accurate and the product promises to be inexpensive. They could take a very small amount of blood and detect cancer anywhere in the body in about five minutes, with a sensitivity that is a thousand times better than in a conventional laboratory test. These devices that are built with nanowires to detect cancer proteins; each nanowire detector is primed to be sensitive to a different cancer marker. The biggest advantage of the nanowire detectors is that they could test for anywhere from ten to one hundred similar medical conditions without adding cost to the testing device. Nanotechnology has also helped to personalize oncology for the detection, diagnosis, and treatment of cancer. It is now able to be tailored to each individual's tumor for better performance. They have found ways that they will be able to target a specific part of the body that is being affected by cancer.

Blood Purification

Magnetic micro particles are proven research instruments for the separation of cells and proteins from complex media. The technology is available under the name Magnetic-activated cell sorting or Dynabeads among others. More recently it was shown in animal models that magnetic nanoparticles can be used for the removal of various noxious compounds including toxins, pathogens, and proteins from whole blood in an extracorporeal circuit similar to dialysis. In contrast to dialysis, which works on the principle of the size related diffusion of solutes and ultrafiltration of fluid across a semi-permeable membrane, the purification with nanoparticles allows specific targeting of substances. Additionally larger compounds which are commonly not dialyzable can be removed.

The purification process is based on functionalized iron oxide or carbon coated metal nanoparticles with ferromagnetic or superparamagnetic properties. Binding agents such as proteins, antibodies, antibiotics, or synthetic ligands are covalently linked to the particle surface. These binding agents are able to interact with target species forming an agglomerate. Applying an external magnetic field gradient allows exerting a force on the nanoparticles. Hence the particles can be separated from the bulk fluid, thereby cleaning it from the contaminants.

The small size (< 100 nm) and large surface area of functionalized nanomagnets leads to advantageous properties compared to hemoperfusion, which is a clinically used technique for the purification of blood and is based on surface adsorption. These advantages are high loading and accessibility of the binding agents, high selectivity towards the target compound, fast diffusion, small hydrodynamic resistance, and low dosage.

This approach offers new therapeutic possibilities for the treatment of systemic infections such as sepsis by directly removing the pathogen. It can also be used to selectively remove cytokines or endotoxins or for the dialysis of compounds which are not accessible by traditional dialysis methods. However the technology is still in a preclinical phase and first clinical trials are not expected before 2017.

Tissue Engineering

Nanotechnology may be used as part of tissue engineering to help reproduce or repair or reshape damaged tissue using suitable nanomaterial-based scaffolds and growth factors. Tissue engineering if successful may replace conventional treatments like organ transplants or artificial implants. Nanoparticles such as graphene, carbon nanotubes, molybdenum disulfide and tungsten disulfide are being used as reinforcing agents to fabricate mechanically strong biodegradable polymeric nanocomposites for bone tissue engineering applications. The addition of these nanoparticles in the polymer matrix at low concentrations (~0.2 weight %) leads to significant improvements in the compressive and flexural mechanical properties of polymeric nanocomposites. Potentially, these nanocomposites may be used as a novel, mechanically strong, light weight composite as bone implants.

For example, a flesh welder was demonstrated to fuse two pieces of chicken meat into a single piece using a suspension of gold-coated nanoshells activated by an infrared laser. This could be used to weld arteries during surgery. Another example is nanonephrology, the use of nanomedicine on the kidney.

Medical Devices

Neuro-electronic interfacing is a visionary goal dealing with the construction of nanodevices that will permit computers to be joined and linked to the nervous system. This idea requires the building of a molecular structure that will permit control and detection of nerve impulses by an external computer. A refuelable strategy implies energy is refilled continuously or periodically with external sonic, chemical, tethered, magnetic, or biological electrical sources, while a nonrefuelable strategy implies that all power is drawn from internal energy storage which would stop when all energy is drained. A nanoscale enzymatic biofuel cell for self-powered nanodevices have been developed that uses glucose from biofluids including human blood and watermelons. One limitation to this innovation is the fact that electrical interference or leakage or overheating from power consumption is possible. The wiring of the structure is extremely difficult because they must be positioned precisely in the nervous system. The structures that will provide the interface must also be compatible with the body's immune system.

Molecular nanotechnology is a speculative subfield of nanotechnology regarding the possibility of engineering molecular assemblers, machines which could re-order matter at a molecular or atomic scale. Nanomedicine would make use of these nanorobots, introduced into the body, to repair or detect damages and infections. Molecular nanotechnology is highly theoretical, seeking to anticipate what inventions nanotechnology might yield and to propose an agenda for future inquiry. The proposed elements of molecular nanotechnology, such as molecular assemblers and nanorobots are far beyond current capabilities. Future advances in nanomedicine could give rise to life extension through the repair of many processes thought to be responsible for aging. K. Eric Drexler, one of the founders of nanotechnology, postulated cell repair machines, including ones operating within cells and utilizing as yet hypothetical molecular machines, in his 1986 book Engines of Creation, with the first technical discussion of medical nanorobots by Robert Freitas appearing in 1999. Raymond Kurzweil, a futurist and transhumanist, stated in his book *The Singularity Is Near* that he believes that advanced medical nanorobotics could completely remedy the effects of aging by 2030.According to Richard Feynman, it was his former graduate student and collaborator Albert Hibbs who originally suggested to him (circa 1959) the idea of a *medical* use for Feynman's theoretical micromachines (see nanotechnology). Hibbs suggested that certain repair machines might one day be reduced in size to the point that it would, in theory, be possible to (as Feynman put it) swallow the doctor. The idea was incorporated into Feynman›s 1959 essay *There's Plenty of Room at the Bottom.*

Potential uses for nanorobotics in medicine include early diagnosis and targeted drug-delivery for cancer, biomedical instrumentation,surgery, pharmacokinetics, monitoring of diabetes, and health care.

In such plans, future medical nanotechnology is expected to employ nanorobots injected into the patient to perform work at a cellular level. Such nanorobots intended for use in medicine should be non-replicating, as replication would needlessly increase device complexity, reduce reliability, and interfere with the medical mission.

Nanotechnology provides a wide range of new technologies for developing customized means to optimize the delivery of pharmaceutical drugs. Today, harmful side effects of treatments such as chemotherapy are commonly a result of drug delivery methods that don't pinpoint their intended target cells accurately. Researchers at Harvard and MIT, however, have been able to attach special RNA strands, measuring nearly 10 nm in diameter, to nanoparticles, filling them with a chemotherapy drug. These RNA strands are attracted to cancer cells. When the nanoparticle encounters a cancer cell, it adheres to it, and releases the drug into the cancer cell. This directed method of drug delivery has great potential for treating cancer patients while avoiding negative effects (commonly associated with improper drug delivery). The first demonstration of nanomotors operating in living organism was carried out in 2014 at University of California, San Diego. MRI-guided nanocapsules are one potential precursor to nanorobots.

Another useful application of nanorobots is assisting in the repair of tissue cells alongside white blood cells. Recruiting inflammatory cells or white blood cells (which include neutrophil granulocytes, lymphocytes, monocytes, and mast cells) to the affected area is the first response of tissues to injury. Because of their small size, nanorobots could attach themselves to the surface of recruited white cells, to squeeze their way out through the walls of blood vessels and arrive at the injury site, where they can assist in the tissue repair process. Certain substances could possibly be used to accelerate the recovery.

The science behind this mechanism is quite complex. Passage of cells across the blood endothelium, a process known as transmigration, is a mechanism involving engagement of cell surface receptors to adhesion molecules, active force exertion and dilation of the vessel walls and physical deformation of the migrating cells. By attaching themselves to migrating inflammatory cells, the robots can in effect "hitch a ride" across the blood vessels, bypassing the need for a complex transmigration mechanism of their own.

As of 2016, in the United States, Food and Drug Administration (FDA) regulates nanotechnology on the basis of size.

REFERENCES

1. Vaughn JR (2006). "Over the Horizon: Potential Impact of Emerging Trends in Information and Communication Technology on Disability Policy and Practice". National Council on Disability, Washington DC: 1–55.
2. Ghosh, A.; Fischer, P. (2009). "Controlled Propulsion of Artificial Magnetic Nanostructured Propellers". Nano Letters. 9 (6): 2243–2245. Bibcode:2009NanoL...9.2243G. doi:10.1021/nl900186w. PMID 19413293.

3. Sierra, D. P.; Weir, N. A.; Jones, J. F. (2005). "A review of research in the field of nanorobotics". U.S. Department of Energy – Office of Scientific and Technical Information Oak Ridge, TN. SAND2005-6808: 1–50. doi:10.2172/875622.
4. Tarakanov, A. O.; Goncharova, L. B.; Tarakanov Y. A. (2009). "Carbon nanotubes towards medicinal biochips". Wiley Interdisciplinary Reviews: Nanomedicine and Nanobiotechnology. 2 (1): 1–10. doi:10.1002/wnan.69.
5. Ignatyev, M. B. (2010). "Necessary and sufficient conditions of nanorobot synthesis". Doklady Mathematics. 82 (1): 671–675. doi:10.1134/S1064562410040435.
6. Cerofolini, G.; Amato, P.; Asserini, M.; Mauri, G. (2010). "A Surveillance System for Early-Stage Diagnosis of Endogenous Diseases by Swarms of Nanobots". Advanced Science Letters. 3 (4): 345–352. doi:10.1166/asl.2010.1138.
7. Yarin, A. L. (2010). "Nanofibers, nanofluidics, nanoparticles and nanobots for drug and protein delivery systems". Scientia Pharmaceutica Central European Symposium on Pharmaceutical Technology. 78 (3): 542. doi:10.3797/scipharm.cespt.8.L02.
8. Wang, J. (2009). "Can Man-Made Nanomachines Compete with Nature Biomotors?". ACS Nano. 3 (1): 4–9. doi:10.1021/nn800829k. PMID 19206241.
9. Amrute-Nayak, M.; Diensthuber, R. P.; Steffen, W.; Kathmann, D.; Hartmann, F. K.; Fedorov, R.; Urbanke, C.; Manstein, D. J.; Brenner, B.; Tsiavaliaris, G. (2010). "Targeted Optimization of a Protein Nanomachine for Operation in Biohybrid Devices". Angewandte Chemie. 122 (2): 322–326. doi:10.1002/ange.200905200.
10. Patel, G. M.; Patel, G. C.; Patel, R. B.; Patel, J. K.; Patel, M. (2006). "Nanorobot: A versatile tool in nanomedicine". Journal of Drug Targeting. 14 (2): 63–67. doi:10.1080/10611860600612862. PMID 16608733.
11. Balasubramanian, S.; Kagan, D.; Jack Hu, C. M.; Campuzano, S.; Lobo-Castañon, M. J.; Lim, N.; Kang, D. Y.; Zimmerman, M.; Zhang, L.; Wang, J. (2011). "Micromachine-Enabled Capture and Isolation of Cancer Cells in Complex Media". Angewandte Chemie International Edition. 50 (18): 4161–4164. doi:10.1002/anie.201100115.
12. Richard P. Feynman (December 1959). "There's Plenty of Room at the Bottom". Retrieved March 2010. Check date values in: |access-date= (help)
13. Zyvex: "Self replication and nanotechnology" "
14. "Foresight Guidelines for Responsible Nanotechnology Development" "
15. Fisher, B. (2008). "Biological Research in the Evolution of Cancer Surgery: A Personal Perspective". Cancer Research. 68 (24): 10007–10020. doi:10.1158/0008-5472.CAN-08-0186. PMID 19074862.
16. Cavalcanti, A.; Shirinzadeh, B.; Zhang, M.; Kretly, L. C. (2008). "Nanorobot Hardware Architecture for Medical Defense". Sensors. 8 (5): 2932–2958. doi:10.3390/s8052932.
17. Hill, C.; Amodeo, A.; Joseph, J. V.; Patel, H. R. (2008). "Nano- and microrobotics: How far is the reality?". Expert Review of Anticancer Therapy. 8 (12): 1891–1897.

doi:10.1586/14737140.8.12.1891. PMID 19046109.

18. Cale, T. S.; Lu, J. Q.; Gutmann, R. J. (2008). "Three-Dimensional Integration in Microelectronics: Motivation, Processing, and Thermomechanical Modeling". Chemical Engineering Communications. 195 (8): 847–888. doi:10.1080/00986440801930302.

19. Couvreur, P.; Vauthier, C. (2006). "Nanotechnology: Intelligent Design to Treat Complex Disease". Pharmaceutical Research. 23 (7): 1417–1450. doi:10.1007/s11095-006-0284-8. PMID 16779701.

20. Elder, J. B.; Hoh, D. J.; Oh, B. C.; Heller, A. C.; Liu, C. Y.; Apuzzo, M. L. J. (2008). "The Future of Cerebral Surgery". Neurosurgery. 62: SHC1555. doi:10.1227/01.neu.0000333820.33143.0d. PMID 18695575.

21. Wong, P. C.; Wong, K. K.; Foote, H. (2003). "Organic data memory using the DNA approach". Communications of the ACM. 46: 95–98. doi:10.1145/602421.602426.

22. Seeman. N. C. (2005). "From genes to machines: DNA nanomechanical devices". Trends in Biochemical Sciences. 30 (3): 119–125. doi:10.1016/j.tibs.2005.01.007. PMC 3471994 . PMID 15752983.

23. Montemagno, C.; Bachand, G. (1999). "Constructing nanomechanical devices powered by biomolecular motors". Nanotechnology. 10 (3): 225–231. Bibcode:1999Nanot..10..225M. doi:10.1088/0957-4484/10/3/301.

24. Yin, P.; Choi, H. M. T.; Calvert, C. R.; Pierce, N. A. (2008). "Programming biomolecular self-assembly pathways". Nature. 451 (7176): 318–322. doi:10.1038/nature06451. PMID 18202654.

25. Douglas, Shawn M.; Bachelet, Ido; Church, George M. (17 February 2012). "A logic-gated nanorobot for targeted transport of molecular payloads". Science. 335 (6070): 831–834. doi:10.1126/science.1214081.

26. Jin, S.; Ye, K. (2007). "Nanoparticle-Mediated Drug Delivery and Gene Therapy". Biotechnology Progress. 23 (1): 32–41. doi:10.1021/bp060348j. PMID 17269667.

27. Hess, Henry; Bachand, George D.; Vogel, Viola (2004). "Powering Nanodevices with Biomolecular Motors". Chemistry A European Journal. 10 (9): 2110–2116. doi:10.1002/chem.200305712.

28. Carroll, G. T.; London, G. B.; Landaluce, T. F. N.; Rudolf, P.; Feringa, B. L. (2011). "Adhesion of Photon-Driven Molecular Motors to Surfacesvia1,3-Dipolar Cycloadditions: Effect of Interfacial Interactions on Molecular Motion". ACS Nano. 5: 622–630. doi:10.1021/nn102876j. PMID 21207983.

29. "Nanofactory Technical Challenges". molecularassembler.com.

30. Martel, S.; Mohammadi, M.; Felfoul, O.; Zhao Lu; Pouponneau, P. (2009). "Flagellated Magnetotactic Bacteria as Controlled MRI-trackable Propulsion and Steering Systems for Medical Nanorobots Operating in the Human Microvasculature". The International Journal of Robotics Research. 28 (4): 571–582. doi:10.1177/0278364908100924. PMC 2772069 . PMID 19890435.

31. Park, S.; Park, S.; Cho, S.; Kim, D.; Lee, Y.; Ko, S.; Hong, Y.; Choy, H.; Min, J.; Park, J.; Park, S. (2013). "New paradigm for tumor theranostic methodology

using bacteria-based microrobot". Scientific Reports. 3: 3394. doi:10.1038/srep03394. PMC 3844944 . PMID 24292152.

32. Sakar, Mahmud (2010). "MicroBioRobots for Single Cell" (PDF).
33. Berry, V.; Saraf, R. F. (2005). "Self-assembly of nanoparticles on live bacterium: An avenue to fabricate electronic devices". Angewandte Chemie International Edition. 44 (41): 6668–6673. doi:10.1002/anie.200501711.
34. RCSB Protein Data Bank. "RCSB PDB-101". rcsb.org.
35. Perkel, Jeffrey M. Viral Mediated Gene Delivery. sciencemag.org
36. Cepko, C.; Pear, W. (2001). "Overview of the Retrovirus Transduction System". Current Protocols in Molecular Biology. doi:10.1002/0471142727.mb0909s36. ISBN 0471142727.
37. Jha, Alok (11 September 2011). "Glow cat: fluorescent green felines could help study of HIV". the Guardian.
38. Cavalcanti, A. (2009). "Nanorobot Invention and Linux: The Open Technology Factor – An Open Letter to UNO General Secretary"(PDF). CANNXS Project. 1 (1): 1–4.
39. Huilgol, N.; Hede, S. (2006). """Nano": The new nemesis of cancer". Journal of Cancer Research and Therapeutics. 2 (4): 186–95. doi:10.4103/0973-1482.29829. PMID 17998702.
40. Das, S.; Gates, A. J.; Abdu, H. A.; Rose, G. S.; Picconatto, C. A.; Ellenbogen, J. C. (2007). "Designs for Ultra-Tiny, Special-Purpose Nanoelectronic Circuits". IEEE Transactions on Circuits and Systems I: Regular Papers. 54 (11): 2528–2540. doi:10.1109/TCSI.2007.907864.
41. Solomon, N., Nanorobotics System, WIPO Patent WO/2008/063473, 2008.
42. Kurzweil, R., Systems and Methods for Generating Biological Material, WIPO Patent WO/2007/001962, 2007.
43. Rosso, F.; Barbarisi, M.; Barbarisi, A. (2011). Technology for Biotechnology. Biotechnology in Surgery. pp. 61–73. doi:10.1007/978-88-470-1658-3_4. ISBN 978-88-470-1657-6.
44. Challacombe, B.; Althoefer, K.; Stoianovici, D. (2010). "Emerging Robotics". New Technologies in Urology. 7: 49–56. doi:10.1007/978-1-84882-178-1_7. ISBN 978-1-84882-177-4.
45. Murday, J. S.; Siegel, R. W.; Stein, J.; Wright, J. F. (2009). "Translational nanomedicine: Status assessment and opportunities". Nanomedicine: Nanotechnology, Biology and Medicine. 5 (3): 251–273. doi:10.1016/j.nano.2009.06.001. PMID 19540359.
46. Hogg, T. (2007). "Coordinating Microscopic Robots in Viscous Fluids". Autonomous Agents and Multi-Agent Systems. 14 (3): 271–305. doi:10.1007/s10458-006-9004-3.
47. Ispir, M., Oktem, L., Method and apparatus for using entropy in ant colony optimization circuit design from high level synthesis, US Patent US8296711 B2, 2010.

48. Ball, H. H., Lucas, M. R., Goutzoulis, A. P. U.S. Patent 7,783,994"Method for providing secure and trusted ASICs using 3D integration", 2010.

49. Pfister, M. U.S. Patent 20,110,048,433 "Method for forming an interventional aid with the aid of self-organizing nanorobots consisting of catoms and associated system unit", 2011.

50. Cuschieri, A. (2005). "Laparoscopic surgery: current status, issues and future developments". Surgeon. 3 (3): 125–138. doi:10.1016/S1479-666X(05)80032-0.

51. Roco, M. C. (2003). "Nanotechnology: convergence with modern biology and medicine". Current Opinion in Biotechnology. 14 (3): 337–346. doi:10.1016/S0958-1669(03)00068-5. PMID 12849790.

52. Scheufele, D. A.; Lewenstein, B. V. (2005). "The Public and Nanotechnology: How Citizens Make Sense of Emerging Technologies". Journal of Nanoparticle Research. 7 (6): 659–667. doi:10.1007/s11051-005-7526-2.

53. Smith, D. M.; Goldstein, D. S.; Heideman, J. (2007). "Reverse Mergers and Nanotechnology". Nanotechnology Law & Business. 4 (3).

54. Morrison, S. (2008). "The Unmanned Voyage: An Examination of Nanorobotic Liability" (PDF). Albany Law Journal of Science & Technology. 18 (229).

55. Craig Tyler, Patent Pirates Search For Texas Treasure, Texas Lawyer, September 20, 2004

56. Jaffe, A. B.; Lerner, J. (2004). Innovation and Its Discontents: How Our Broken Patent System is Endangering Innovation and Progress, and What to Do About It. ISBN 0-691-11725-X.

57. Gilbert, R. J.; Newbery, D. M. G. (June 1982). "Preemptive Patenting and the Persistence of Monopoly". American Economic Review. 72 (3): 514–526. JSTOR 1831552.

58. Nano Robot by 3D Printing (Seoul National University, Korea).wmv, 2012-01-29, retrieved 2015-12-04

59. "Nanotechnology and 3D-printing". www.nanowerk.com. Retrieved 2015-12-04.

60. Nanotechnology in Cancer. nano.cancer.gov

61. Zyga, Lisa (December 5, 2007) "Virtual 3D nanorobots could lead to real cancer-fighting technology". physorg.com.

62. Lavan, D. A.; McGuire, T.; Langer, R. (2003). "Small-scale systems for in vivo drug delivery". Nature Biotechnology. 21 (10): 1184–91. doi:10.1038/nbt876. PMID 14520404.

63. "(Emerging Technologies) Software Provides Peek into the Body—and the Future (MPMN archive, March 08)". nanorobotdesign.com.

64. Leary, S. P.; Liu, C. Y.; Apuzzo, M. L. J. (2006). "Toward the Emergence of Nanoneurosurgery: Part III???Nanomedicine: Targeted Nanotherapy, Nanosurgery, and Progress Toward the Realization of Nanoneurosurgery". Neurosurgery. 58 (6): 1009–1026. doi:10.1227/01.NEU.0000217016.79256.16. PMID 16723880.

65. Tiny robot useful for surgery

66. Shanthi, Vadali; Sravani Musunuri (13 November 2007). "Prospects for Medical Robots". AZoNano. doi:10.2240/azojono0119.
67. Melki, Benjamin (January 31, 2007) Nanorobotics for Diabetes. nanovip.com
68. Donnelly, R. (2007). "Wellness engineering and health management: A video interview with Harold H. Szu". SPIE Newsroom. doi:10.1117/2.3200708.0002.
69. Bhowmik, Debjit (2009). "Role of Nanotechnology in novel Drug Delivery system" (PDF). Journal of Pharmaceutical Science and Technology. 1 (1): 20–35.
70. Bullis, Kevin (April 29, 2008). "Nano RNA Delivery." MIT Technology Review.
71. Gao, W.; Wang, J. (2014). "Synthetic micro/nanomotors in drug delivery". Nanoscale. 6 (18): 10486–94. Bibcode:2014Nanos...610486G. doi:10.1039/C4NR03124E. PMID 25096021.
72. Gao, W.; Dong, R.; Thamphiwatana, S.; Li, J.; Gao, W.; Zhang, L.; Wang, J. (2015). "Artificial Micromotors in the Mouse's Stomach: A Step towardin Vivo Use of Synthetic Motors". ACS Nano. 9 (1): 117–23. doi:10.1021/nn507097k. PMC 4310033 . PMID 25549040.
73. Vartholomeos, P.; Fruchard, M.; Ferreira, A.; Mavroidis, C. (2011). "MRI-Guided Nanorobotic Systems for Therapeutic and Diagnostic Applications". Annu Rev Biomed Eng. 13: 157–84. doi:10.1146/annurev-bioeng-071910-124724.
74. Casal, Arancha et al. (2004) "Nanorobots As Cellular Assistants in Inflammatory Responses". nanorobotdesign.com
75. C. Janeway (ed.) (2001) ImmunoBiology, the Immune System in Health and Disease. Garland Pub; 5th ed. ISBN 0-8153-3642-X.
76. FDA (2011) Considering Whether an FDA-Regulated Product Involves the Application of Nanotechnology, Guidance for Industry, Draft Guidance.
77. Excell, Jon. "The rise of additive manufacturing". The Engineer. Retrieved 2013-10-30.
78. "3D Printer Technology – Animation of layering". Create It Real. Retrieved 2012-01-31.
79. "Jeremy Rifkin and The Third Industrial Revolution Home Page". Thethirdindustrialrevolution.com. Retrieved 2016-01-04.
80. "A third industrial revolution". The Economist. 2012-04-21. Retrieved 2016-01-04.
81. Standard Terminology for Additive Manufacturing – General Principles – Terminology. ASTM International. September 2013, Retrieved 2016-07-11
82. Jane Bird (2012-08-08). "Exploring the 3D printing opportunity". The Financial Times. Retrieved 2012-08-30.
83. Hideo Kodama, "A Scheme for Three-Dimensional Display by Automatic Fabrication of Three-Dimensional Model," IEICE TRANSACTIONS on Electronics (Japanese Edition), vol.J64-C, No.4, pp.237–241, April 1981
84. Hideo Kodama, "Automatic method for fabricating a three-dimensional plastic model with photo-hardening polymer," Review of Scientific Instruments, Vol. 52, No. 11, pp. 1770–1773, November 1981

85. Jean-Claude, Andre. "Disdpositif pour realiser un modele de piece industrielle". National De La Propriete Industrielle.

86. Mendoza, Hannah Rose (2015-05-15). "Alain Le Méhauté, The Man Who Submitted Patent For SLA 3D Printing Before Chuck Hull". 3dprint.com.

87. Moussion, Alexandre (2014). "Interview d'Alain Le Méhauté, l'un des pères de l'impression 3D". Primante 3D.

88. "3D Printing: What You Need to Know". PCMag.com. Retrieved 2013-10-30.

89. Apparatus for Production of Three-Dimensional Objects by Stereolithography (1984-08-08)

90. Freedman, David H (2012). "Layer By Layer". Technology Review. 115 (1): 50–53.

91. Amon, C. H.; Beuth, J. L.; Weiss, L. E.; Merz, R.; Prinz, F. B. (1998). "Shape Deposition Manufacturing With Microcasting: Processing, Thermal and Mechanical Issues" (PDF). Journal of Manufacturing Science and Engineering. 120 (3). Retrieved 2014-12-20.

92. Beck, J.E.; Fritz, B.; Siewiorek, Daniel; Weiss, Lee (1992). "Manufacturing Mechatronics Using Thermal Spray Shape Deposition" (PDF). Proceedings of the 1992 Solid Freeform Fabrication Symposium. Retrieved 2014-12-20.

93. Prinz, F. B.; Merz, R.; Weiss, Lee (1997). Ikawa, N., ed. Building Parts You Could Not Build Before. Proceedings of the 8th International Conference on Production Engineering. 2-6 Boundary Row, London SE1 8HN, UK: Chapman & Hall. pp. 40–44.

94. "Google Ngram of the term additive manufacturing".

95. GrabCAD, GE jet engine bracket challenge

96. Zelinski, Peter (2014-06-02), "How do you make a howitzer less heavy?", Modern Machine Shop

97. Pearce, Joshua M.; et al. (2010). "3-D Printing of Open Source Appropriate Technologies for Self-Directed Sustainable Development". Journal of Sustainable Development. 3 (4): 17–29. doi:10.5539/jsd.v3n4p17. Retrieved 2012-01-31.

98. "3D Printing". 13 February 2012.

99. Ishengoma, Fredrick R.; Mtaho, Adam B. (2014-10-18). "3D Printing: Developing Countries Perspectives". International Journal of Computer Applications. 104 (11): 30–34. arXiv:1410.5349 . doi:10.5120/18249-9329. ISSN 0975-8887.

100. Jacobs, Paul Francis (1992-01-01). Rapid Prototyping & Manufacturing: Fundamentals of Stereolithography. Society of Manufacturing Engineers. ISBN 978-0-87263-425-1.

101. "3D solid repair software – Fix STL polygon mesh files – LimitState:FIX". Print.limitstate.com. Retrieved 2016-01-04.

102. "3D Printing Pens". yellowgurl.com. Retrieved 2016-08-09.

103. "Model Repair Service". Modelrepair.azurewebsites.net. Retrieved 2016-01-04.

104. "Magics, the Most Powerful 3D Printing Software | Software for additive

manufacturing". Software.materialise.com. Retrieved 2016-01-04.

105. "netfabb Cloud Services". Netfabb.com. 2009-05-15. Retrieved 2016-01-04.
106. "How to repair a 3D scan for printing". Anamarva.com. Retrieved 2016-01-04.
107. Fausto Bernardini, Holly E. Rushmeier (2002). "The 3D Model Acquisition Pipeline GAS" (PDF). Comput. Graph. Forum. 21 (2): 149–172. doi:10.1111/1467-8659.00574.
108. "cord or cordless 3D printing pen". tutorial45. Retrieved 3 June 2016.
109. "Objet Connex 3D Printers". Objet Printer Solutions. Retrieved 2012-01-31.
110. "How to 3D-print super-fast and have an awesome finishing". 3dprinterchat. Retrieved 5 May 2016.
111. "How to Smooth 3D-Printed Parts".
112. Kraft, Caleb. "Smoothing Out Your 3D Prints With Acetone Vapor". Make. Make. Retrieved 2016-01-05.
113. Amberlee S. Haselhuhn, Eli J. Gooding, Alexandra G. Glover, Gerald C. Anzalone, Bas Wijnen, Paul G. Sanders, Joshua M. Pearce. (2014). "Substrate Release Mechanisms for Gas Metal Arc 3-D Aluminum Metal Printing". 3D Printing and Additive Manufacturing. 1 (4): 204–209. doi:10.1089/3dp.2014.0015.
114. Haselhuhn, Amberlee S.; Wijnen, Bas; Anzalone, Gerald C.; Sanders, Paul G.; Pearce, Joshua M. (2015). "In Situ Formation of Substrate Release Mechanisms for Gas Metal Arc Weld Metal 3-D Printing". Journal of Materials Processing Technology. 226: 50–59. doi:10.1016/j.jmatprotec.2015.06.038.
115. Sames, W. "The metallurgy and processing science of metal additive manufacturing". International Materials Reviews. 61: 315–360. doi:10.1080/09506608.2015.1116649.
116. FDM is a proprietary term owned by Stratasys. All 3-D printers that are not Stratasys machines and use a fused filament process are referred to as or fused filament fabrication (FFF).
117. Sherman, Lilli Manolis (November 15, 2007). "A whole new dimension – Rich homes can afford 3D printers". The Economist.
118. Wohlers, Terry. "Factors to Consider When Choosing a 3D Printer (WohlersAssociates.com, Nov/Dec 2005)".
119. www.3ders.org (2012-09-25). "Casting aluminum parts directly from 3D printed PLA parts". 3ders.org. Retrieved 2013-10-30.
120. "Affordable 3D Printing with new Selective Heat Sintering (SHS™) technology". blueprinter.
121. Chee Kai Chua; Kah Fai Leong; Chu Sing Lim (2003). Rapid Prototyping. World Scientific. p. 124. ISBN 978-981-238-117-0.
122. "PTFE Tubing Benefits & Applications". Fluorotherm Polymers, Inc. Retrieved 17 November 2015.
123. "Aluminum-powder DMLS-printed part finishes race first".

124. "Selective Laser Sintering, Birth of an Industry".
125. Deckard, C., "Method and apparatus for producing parts by selective sintering", U.S. Patent 4,863,538, filed October 17, 1986, published September 5, 1989.
126. Housholder, R., "Molding Process", U.S. Patent 4,247,508, filed December 3, 1979, published January 27, 1981.
127. Hiemenz, Joe. "Rapid prototypes move to metal components (EE Times, 3/9/2007)".
128. "Rapid Manufacturing by Electron Beam Melting". SMU.edu.
129. "3D Printer Uses Standard Paper".
130. U.S. Patent 4,575,330
131. "Rapid Prototyping in Europe and Japan" (PDF). JTEC/WTEC Panel Report. Retrieved 28 December 2016.
132. Beaumont Newhall (May 1958) "Photosculpture," Image, 7 (5) : 100–105
133. François Willème, "Photo-sculpture," U.S. Patent no. 43,822 (August 9, 1864). Available on-line at: U.S. Patent 43,822
134. François Willème (May 15, 1861) "La sculpture photographique", Le Moniteur de la photographie, p. 34.
135. "EnvisionTEC Perfactory". EnvisionTEC.
136. Johnson, R. Colin. "Cheaper avenue to 65 nm? (EE Times, 3/30/2007)".
137. "The World's Smallest 3D Printer". TU Wien. 12 September 2011.
138. "3D-printing multi-material objects in minutes instead of hours". Kurzweil Accelerating Intelligence. November 22, 2013.
139. St. Fleur, Nicholas (17 March 2015). "3-D Printing Just Got 100 Times Faster". The Atlantic. Retrieved 19 March 2015.
140. Castelvecchi, Davide (17 March 2015). "Chemical trick speeds up 3D printing". Nature. Retrieved 19 March 2015.
141. Saxena, Shalini (19 March 2015). "New nonstop 3D printing process takes only minutes instead of hours". Ars Technica. Retrieved 19 March 2015.
142. Beese, Allison M.; Carroll, Beth E. (2015-12-21). "Review of Mechanical Properties of Ti-6Al-4V Made by Laser-Based Additive Manufacturing Using Powder Feedstock". JOM. 68 (3): 724–734. doi:10.1007/s11837-015-1759-z. ISSN 1047-4838.
143. Gibson, Ian; Rosen, David; Stucker, Brent. "Chapter 10". Additive Manufacturing Technologies - Springer. doi:10.1007/978-1-4939-2113-3.
144. Yuenyong Nilsiam, Amberlee Haselhuhn, Bas Wijnen, Paul Sanders, & Joshua M. Pearce. Integrated Voltage – Current Monitoring and Control of Gas Metal Arc Weld Magnetic Ball-Jointed Open Source 3-D Printer. Machines 3(4), 339-351 (2015).
145. A. Pinar, B. Wijnen, G. C. Anzalone, T. C. Havens, P. G. Sanders, J. M. Pearce. Low-cost Open-Source Voltage and Current Monitor for Gas Metal Arc Weld

3-D Printing. Journal of Sensors Vol. 2015, Article ID 876714, 8 pages, 2015.

146. "3D Printing: Challenges and Opportunities for International Relations". Council on Foreign Relations. October 23, 2013. Retrieved 2013-10-30.
147. "Despite Market Woes, 3D Printing Has a Future Thanks to Higher Education - Bold". 2 December 2015.
148. "Stratasys Ltd. Short Interest Update". Americantradejournal.com.
149. "First MakerBot 3D Printing Innovation Center in the Midwest opens at Xavier".
150. "UMass Amherst Library Opens 3-D Printing Innovation Center".
151. Kalish, Jon. "A Space For DIY People To Do Their Business (NPR.org, November 28, 2010)". Retrieved 2012-01-31.
152. Jones, R.; Haufe, P.; Sells, E.; Iravani, P.; Olliver, V.; Palmer, C.; Bowyer, A. (2011). "Reprap-- the replicating rapid prototyper". Robotica. 29 (1): 177–191. doi:10.1017/s026357471000069x.
153. "Open source 3D printer copies itself". Computerworld New Zealand. 2008-04-07. Retrieved 2013-10-30.
154. "First reprapped circuit". RepRap.
155. "An Inexpensive Way to Print Out Metal Parts". The New York Times. 10 December 2013.
156. Anzalone, Gerald C.; Zhang, Chenlong; Wijnen, Bas; Sanders, Paul G.; Pearce, Joshua M. (2013). "Low-Cost Open-Source 3-D Metal Printing". IEEE Access. 1: 803–810. doi:10.1109/ACCESS.2013.2293018.
157. "3D Printing Trends July 2016". 3D Hubs. Retrieved 2016-07-12.
158. "What's up with Original Prusa i3? - Prusa Printers". 2016-03-02. Retrieved 2016-07-12.
159. Bilton, Nick. "Disruptions: On the Fast Track to Routine 3-D Printing".
160. www.3ders.org. "3D printers list with prices". 3ders.org. Retrieved 2013-10-30.
161. Simonite, Tom. "Desktop fabricator may kick-start home revolution".
162. "Rapide One – Affordable Professional Desktop 3D Printer by Rapide 3D". Indiegogo. December 2, 2013. Retrieved 20 January 2014.
163. "A Review Of The 3Doodler Pen, Which Raised Over $2 Million On Kickstarter".
164. Dorrier, Jason (27 February 2013). "Kickstarter 3Doodler 3D Printing Pen Nothing of the Sort - But Somehow Raises $2 Million".
165. Wittbrodt, B. T.; Glover, A. G.; Laureto, J.; Anzalone, G. C.; Oppliger, D.; Irwin, J. L.; Pearce, J. M. (2013). "Life-cycle economic analysis of distributed manufacturing with open-source 3-D printers". Mechatronics. 23 (6): 713–726. doi:10.1016/j.mechatronics.2013.06.002.
166. Kreiger, M.; Pearce, J. M. (2013). "Environmental Life Cycle Analysis of Distributed Three-Dimensional Printing and Conventional Manufacturing of Polymer Products". ACS Sustainable Chemistry & Engineering. 1: 131002082320002. doi:10.1021/sc400093k.

167. Baechler, Christian; DeVuono, Matthew; Pearce, Joshua M. (2013). "Distributed Recycling of Waste Polymer into RepRap Feedstock". Rapid Prototyping Journal. 19 (2): 118–125. doi:10.1108/13552541311302978.

168. "Distributed Recycling of Post-Consumer Plastic Waste in Rural Areas".

169. "Rostock (delta robot 3D printer) by Johann". Thingiverse.com.

170. Vandendriessche, Pieter-Jan. "delta 3D printer accuracy".

171. Titsch, Mike (July 11, 2013). "MatterHackers Opens 3D Printing Store and Releases MatterControl 0.7.6". Retrieved November 30, 2013.

172. "Hoosier Daddy – The Largest Delta 3D Printer In the World". 3D Printer World. Punchbowl Media. 2014-09-23. Retrieved 2014-09-28.

173. McKenna, Beth (2014-09-26). "The Next Big Thing in 3-D Printing: Big Area Additive Manufacturing, or BAAM". The Motley Fool. Retrieved 2014-09-28.

174. J. L. Wasserman; et al. (2008). "Fabrication of One-Dimensional Programmable-Height Nanostructures via Dynamic Stencil Deposition". Review of Scientific Instruments. 79 (7): 073909. arXiv:0802.1848 . doi:10.1063/1.2960573.

175. patel, Prachi (5 March 2013). "Micro 3-D Printer Creates Tiny Structures in Seconds". MIT Technology Review.

176. "Print me a Stradivarius – How a new manufacturing technology will change the world". Economist Technology. 2011-02-10. Retrieved 2012-01-31.

177. Zelinski, Peter (2014-06-25). "Video: World's largest additive metal manufacturing plant". Modern Machine Shop.

178. Sherman, Lilli Manolis. "3D Printers Lead Growth of Rapid Prototyping (Plastics Technology, August 2004)". Retrieved 2012-01-31.

179. "3D printing: 3D printing scales up". The Economist. 2013-09-07. Retrieved 2013-10-30.

180. "A printed smile". The Economist. ISSN 0013-0613. Retrieved 2016-05-08.

181. "Development of a Three-Dimensional Printed, Liquid-Cooled Nozzle for a Hybrid Rocket Motor," Nick Quigley and James Evans Lyne, Journal of Propulsion and Power, Vol. 30, No. 6 (2014), pp. 1726–1727.

182. Vincent & Earls 2011

183. Anzalone, G.; Wijnen, B.; Pearce, Joshua M. (2015). "Multi-material additive and subtractive prosumer digital fabrication with a free and open-source convertible delta RepRap 3-D printer". Rapid Prototyping Journal. 21 (5): 506–519. doi:10.1108/RPJ-09-2014-0113.

184. Felix Bopp (2010). Future Business Models by Additive Manufacturing. Verlag. ISBN 3-8366-8508-6. Retrieved 4 July 2014.

185. Wu, D.; Thames, J.L.; Rosen, D.W.; Schaefer, D. (2013). "Enhancing the Product Realization Process with Cloud-Based Design and Manufacturing Systems." Transactions of the ASME". Journal of Computing and Information Science in Engineering. 13 (4): 041004. doi:10.1115/1.4025257.

186. Wu, D.; Rosen, D.W.; Wang, L.; Schaefer, D. (2015). "Cloud-Based Design

and Manufacturing: A New Paradigm in Digital Manufacturing and Design Innovation". Computer-Aided Design. 59 (1): 1–14. doi:10.1016/j.cad.2014.07.006.

187. Wu, D.; Rosen, D.W.; Schaefer, D. (2015). "Scalability Planning for Cloud-Based Manufacturing Systems." Transactions of the ASME". Journal of Manufacturing Science and Engineering. 137: 040911. doi:10.1115/1.4030266.
188. "3D Hubs: Like Airbnb For 3D Printers". gizmodo. Retrieved 2014-07-05.
189. Sterling, Bruce (2011-06-27). "Spime Watch: Dassault Systèmes' 3DVIA and Sculpteo (Reuters, June 27, 2011)". Wired. Archived from the original on 15 April 2014. Retrieved 2012-01-31.
190. Vance, Ashlee (2011-01-12). "The Wow Factor of 3-D Printing (The New York Times, January 12, 2011)". Retrieved 2012-01-31.
191. "The action doll you designed, made real". makie.me. Retrieved January 18, 2013.
192. "Cubify — Express Yourself in 3D". myrobotnation.com. Retrieved 2014-01-25.
193. "Turn Your Baby's Cry Into an iPhone Case". Bloomberg Businessweek. 2012-03-10. Retrieved 2013-02-20.
194. "Nokia backs 3D printing for mobile phone cases". BBC News Online. 2013-02-18. Retrieved 2013-02-20.
195. Wohlers Report 2009, State of the Industry Annual Worldwide Progress Report on Additive Manufacturing, Wohlers Associates, ISBN 978-0-9754429-5-1
196. Hopkinson, N & Dickens, P 2006, 'Emerging Rapid Manufacturing Processes', in Rapid Manufacturing; An industrial revolution for the digital age, Wiley & Sons Ltd, Chichester, W. Sussex
197. Symes, M. D.; Kitson, P. J.; Yan, J.; Richmond, C. J.; Cooper, G. J. T.; Bowman, R. W.; Vilbrandt, T.; Cronin, L. (2012). "Integrated 3D-printed reactionware for chemical synthesis and analysis". Nature Chemistry. 4 (5): 349–354. doi:10.1038/nchem.1313. PMID 22522253.
198. Wong, Venessa. "A Guide to All the Food That's Fit to 3D Print (So Far)". Bloomberg.com.
199. "Did BeeHex Just Hit 'Print' to Make Pizza at Home?". Retrieved 28 May 2016.
200. "Foodini 3D Printer Cooks Up Meals Like the Star Trek Food Replicator". Retrieved 27 January 2015.
201. "3D Printing: Food in Space". NASA. Retrieved 2015-09-30.
202. "Print your own medicine".
203. "RFA-HD-15-023: Use of 3-D Printers for the Production of Medical Devices (R43/R44)". NIH grants. Retrieved 2015-09-30.
204. "7 Ways 3D Printing Is Disrupting The Medical Industry". 3D Masterminds.
205. "3D Microporous Scaffolds Manufactured via Combination of Fused Filament Fabrication and Direct Laser Writing Ablation". MDPI. Retrieved 2016-08-24.
206. "3D Printed Clothing Becoming a Reality". Resins Online. 2013-06-17. Retrieved 2013-10-30.

207. Michael Fitzgerald (2013-05-28). "With 3-D Printing, the Shoe Really Fits". MIT Sloan Management Review. Retrieved 2013-10-30.
208. Sharma, Rakesh (2013-09-10). "3D Custom Eyewear The Next Focal Point For 3D Printing". Forbes.com. Retrieved 2013-09-10.
209. Parker C. J. (2015). The Human Acceptance of 3D Printing in Fashion Paradox: Is mass customisation a bridge too far? IWAMA 2015: 5th International Workshop of Advanced Manufacturing and Automation. Shanghai, China.
210. "Karl Lagerfeld Showcases 3D Printed Chanel at Paris Fashion Week".
211. "Noa Ra[illegible] uses grid patterns and 3D printing in fashion collection". 21 August 2014.
212. Holgate, Mark. "Meet Iris van Herpen, the Dutch Designer Boldly Going Into the Future".
213. "Koenigsegg One:1 Comes With 3D Printed Parts". Business Insider. Retrieved 2014-05-14.
214. tecmundo.com.br/ Conheça o Urbee, primeiro carro a ser fabricado com uma impressora 3D
215. Eternity, Max. "The Urbee 3D-Printed Car: Coast to Coast on 10 Gallons?".
216. 3D Printed Car Creator Discusses Future of the Urbee on YouTube
217. "Local Motors shows Strati, the world's first 3D-printed car". 13 January 2015.
218. Walker, Daniela. "Local Motors wants to 3D-print your next car out of plastic".
219. Warren, Tamara (16 June 2016). "This autonomous, 3D-printed bus starts giving rides in Washington, DC today".
220. "Building Olli: Why "Second-degree DDM" is critical to the process - Local Motors". 24 June 2016.
221. Simmons, Dan (2015-05-06). "Airbus had 1,000 parts 3D printed to meet deadline". BBC. Retrieved 2015-11-27.
222. Zitun, Yoav (2015-07-27). "The 3D printer revolution comes to the IAF". Ynet News. Retrieved 2015-09-29.
223. Greenberg, Andy (2012-08-23). "'Wiki Weapon Project' Aims To Create A Gun Anyone Can 3D-Print At Home". Forbes. Retrieved 2012-08-27.
224. Poeter, Damon (2012-08-24). "Could a 'Printable Gun' Change the World?". PC Magazine. Retrieved 2012-08-27.
225. "Blueprints for 3-D printer gun pulled off website". statesman.com. May 2013. Retrieved 2013-10-30.
226. Samsel, Aaron. "3D Printers, Meet Othermill: A CNC machine for your home office (VIDEO)". Guns.com. Retrieved 2013-10-30.
227. "The Third Wave, CNC, Stereolithography, and the end of gun control". Popehat. Retrieved 2013-10-30.
228. Rosenwald, Michael S. (2013-02-25). "Weapons made with 3-D printers could test gun-control efforts". Washington Post.

229. "Making guns at home: Ready, print, fire". The Economist. 2013-02-16. Retrieved 2013-10-30.

230. Rayner, Alex (6 May 2013). "3D-printable guns are just the start, says Cody Wilson". The Guardian. London.

231. Manjoo, Farhad (2013-05-08). "3-D-printed gun: Yes, it will be possible to make weapons with 3-D printers. No, that doesn't make gun control futile". Slate.com. Retrieved 2013-10-30.

232. Franzen, Carl. "3D-printed gun maker in Japan sentenced to two years in prison". The Verge.

233. "Transplant jaw made by 3D printer claimed as first". BBC. 2012-02-06.

234. Rob Stein (2013-03-17). "Doctors Use 3-D Printing To Help A Baby Breathe". NPR.

235. Moore, Calen (11 February 2014). "Surgeons have implanted a 3-D-printed pelvis into a U.K. cancer patient". fiercemedicaldevices.com. Retrieved 2014-03-04.

236. Keith Perry (2014-03-12). "Man makes surgical history after having his shattered face rebuilt using 3D printed parts". London: The Daily Telegraph. Retrieved 2014-03-12.

237. BBC News (October 2014). "Inverness girl Hayley Fraser gets 3D-printed hand", BBC News, 2014-10-01. Retrieved 2014-10-02.

238. "3D-Printed Foot Lets Crippled Duck Walk Again".

239. Pleasance, Chris (18 August 2014). "Puppy power: Chihuahua born without front legs is given turbo-charged makeover after being fitted with 3D printed body harness and a set of skateboard wheels". The Daily Mail. Retrieved 2014-08-21.

240. Flaherty, Joseph (2013-07-30). "So Cute: Hermit Crabs Strut in Stylish 3-D Printed Shells". Wired.

241. "3D Systems preps for global launch of 'printed' knee implants for dogs". FierceAnimalHealth.com. Retrieved 13 April 2015.

242. Saxena, Varun. "FDA clears 3-D printed device for minimally invasive foot surgery". FierceMedicalDevices.com. Retrieved 14 April 2015.

243. Yue, J; Zhao, P; Gerasimov, JY; de Lagemaat, M; Grotenhuis, A; Rustema-Abbing, M; van der Mei, HC; Busscher, HJ; Herrmann, A; Ren, Y (2015). "3D-Printable Antimicrobial Composite Resins". Adv. Funct. Mater. 25 (43): 6756–6767. doi:10.1002/adfm.201502384.

244. "Mish's Global Economic Trend Analysis: 3D-Printing Spare Human Parts; Ears and Jaws Already, Livers Coming Up ; Need an Organ? Just Print It". Globaleconomicanalysis.blogspot.co.uk. 2013-08-18. Retrieved 2013-10-30.

245. Aias, L (11 Aug 2016). "Grecia, the toucan with the prosthetic beak, now receiving visitors". The Tico Times. Retrieved 14 Sep 2016.

246. "3D-printed sugar network to help grow artificial liver". BBC News.

247. "Invetech helps bring bio-printers to life". Australian Life Scientist. Westwick-Farrow Media. December 11, 2009. Retrieved December 31, 2013.

248. "Building body parts with 3D printing".
249. Silverstein, Jonathan. "'Organ Printing' Could Drastically Change Medicine (ABC News, 2006)". Retrieved 2012-01-31.
250. "Engineering Ourselves – The Future Potential Power of 3D-Bioprinting?". ENGINEERING.com.
251. The Diplomat (2013-08-15). "Chinese Scientists Are 3D Printing Ears and Livers – With Living Tissue". Tech Biz. The Diplomat. Retrieved 2013-10-30.
252. "How do they 3D print kidney in China". Retrieved 2013-10-30.
253. Palmer, Eric (3 August 2015). "Company builds plant for 3DP pill making as it nails first FDA approval". fiercepharmamanufacturing.com. Retrieved 4 August 2015.
254. Kuehn, Steven E. (September 2015). "I'm Printing Your Prescription Now, Ma'am". From the Editor. Pharmaceutical Manufacturing (paper). Putnam Media: 7.
255. "The Almost Completely Open Source Laptop Goes on Sale".
256. McCue, TJ. "Robots And 3D Printing".
257. Why to Use 3D Printers and the Best 3D Printers To Build Your Own Robot
258. Printoo: Giving Life to Everyday Objects (paper-thin, flexible Arduino-compatible modules)
259. 3&DBot: An Arduino 3D printer-robot with wheels
260. "A lesson in building a custom 3D printed humanoid robot".
261. "New horizons open with space-based 3D printing". SPIE Newsroom. Retrieved 1 April 2015.
262. Hays, Brooks (2014-12-19). "NASA just emailed the space station a new socket wrench". Retrieved 2014-12-20.
263. Brabaw, Kasandra (2015-01-30). "Europe's 1st Zero-Gravity 3D Printer Headed for Space". Retrieved 2015-02-01.
264. Wood, Anthony (2014-11-17). "POP3D to be Europe's first 3D printer in space". Retrieved 2015-02-01.
265. Raval, Siddharth (2013-03-29). "SinterHab: A Moon Base Concept from Sintered 3D-Printed Lunar Dust". Space Safety Magazine. Retrieved 2013-10-15.
266. "The World's First 3D-Printed Building Will Arrive In 2014". TechCrunch. 2012-01-20. Retrieved 2013-02-08.
267. Diaz, Jesus (2013-01-31). "This Is What the First Lunar Base Could Really Look Like". Gizmodo. Retrieved 2013-02-01.
268. "The RepRap's Heritage".
269. Kelly, Heather (July 31, 2013). "Study: At-home 3D printing could save consumers "thousands"". CNN.
270. Séquin, C. H. (2005). "Rapid prototyping". Communications of the ACM. 48 (6): 66. doi:10.1145/1064830.1064860.

271. Guth, Robert A. "How 3-D Printing Figures To Turn Web Worlds Real (The Wall Street Journal, December 12, 2007)" (PDF). Retrieved 2012-01-31.

272. iPad iPhone Android TIME TV Populist The Page (2008-04-03). "Bathsheba Grossman's Quin.MGX for Materialise". Time. Retrieved 2013-10-30.

273. Williams, Holly (2011-08-28). "Object lesson: How the world of decorative art is being revolutionised by 3D printing (The Independent, 28 August 2011)". London. Retrieved 2012-01-31.

274. Bennett, Neil (November 13, 2013). "How 3D printing is helping doctors mend you better". TechAdvisor.

275. "Custom Bobbleheads". Retrieved 13 January 2015.

276. "3D-print your face in chocolate for that special Valentine's Day gift". The Guardian. 25 January 2013.

277. Cignoni, P.; Scopigno, R. (2008). "Sampled 3D models for CH applications". Journal on Computing and Cultural Heritage. 1: 1–23. doi:10.1145/1367080.1367082.

278. DiSalvo, C (2009). "Design and the Construction of Publics". Design Issues. 1. 25: 48–63. doi:10.1162/desi.2009.25.1.48.

279. Ratto, M. & Ree, R. (2012). "Materializing information: 3D printing and social change.". First Monday. 17 (7). doi:10.5210/fm.v17i7.3968.

280. Ratto, Matt (2011). "Open Design and Critical Making". Open Design Now: Why Design Cannot Remain Exclusive.

281. Lukens, Jonathan. "Speculative Design and Technological Fluency". International Journal of Learning and Media. 3: 23–39. doi:10.1162/ijlm_a_00080.

282. Pandey, S.; Gupta, B.; Nahata, A. (2013). "Complex Geometry Plasmonic Terahertz Waveguides Created via 3D Printing". Cleo: 2013. pp. CTh1K.CTh12. doi:10.1364/CLEO_SI.2013.CTh1K.2. ISBN 978-1-55752-972-5.

283. "I have been working on my #solidsounds... - Janine Ling Carr - Facebook".

284. Fletcher, JC (August 28, 2008). "Virtually Overlooked: Mario Artist". Retrieved 2014-06-14.

285. "Mario Artist: Polygon Studio". Retrieved 2014-06-14.

286. ewilhelm. "3D printed clock and gears". Instructables.com. Retrieved 2013-10-30.

287. 23/01/2012 (2012-01-23). "Successful Sumpod 3D printing of a herringbone gear". 3d-printer-kit.com. Retrieved 2013-10-30.

288. ""backscratcher" 3D Models to Print - yeggi".

289. Sanderson, Katharine. "Make your own drugs with a 3D printer".

290. Cronin, Lee (2012-04-17). "3D printer developed for drugs" (video interview 5:21). Glasgow University: BBC News Online. Retrieved 2013-03-06.

291. D'Aveni, Richard (March 2013). "3-D Printing Will Change the World". Harvard Business Review. Retrieved 2014-10-08.

292. "3D printable SLR brings whole new meaning to "digital camera"". Gizmag.com. Retrieved 2013-10-30.

293. Schelly, C., Anzalone, G., Wijnen, B., & Pearce, J. M. (2015). "Open-source 3-D printing Technologies for education: Bringing Additive Manufacturing to the Classroom." Journal of Visual Languages & Computing.

294. Grujović, N., Radović, M., Kanjevac, V., Borota, J., Grujović, G., & Divac, D. (2011, September). "3D printing technology in education environment." In 34th International Conference on Production Engineering (pp. 29–30).

295. Mercuri, R., & Meredith, K. (2014, March). "An educational venture into 3D Printing." In Integrated STEM Education Conference (ISEC), 2014 IEEE (pp. 1–6). IEEE.

296. Students Use 3D Printing to Reconstruct Dinosaurs on YouTube

297. Gonzalez-Gomez, J., Valero-Gomez, A., Prieto-Moreno, A., & Abderrahim, M. (2012). "A new open source 3d-printable mobile robotic platform for education." In Advances in Autonomous Mini Robots (pp. 49–62). Springer Berlin Heidelberg.

298. J. Irwin, J.M. Pearce, D. Opplinger, and G. Anzalone. The RepRap 3-D Printer Revolution in STEM Education, 121st ASEE Annual Conference and Exposition, Indianapolis, IN. Paper ID #8696 (2014).

299. Zhang, C.; Anzalone, N. C.; Faria, R. P.; Pearce, J. M. (2013). De Brevern, Alexandre G, ed. "Open-Source 3D-Printable Optics Equipment". PLoS ONE. 8 (3): e59840. doi:10.1371/journal.pone.0059840. PMC 3609802 . PMID 23544104.

300. "3D Printing in the Classroom to Accelerate Adoption of Technology".

301. Kostakis, V.; Niaros, V.; Giotitsas, C. (2014). "Open source 3D printing as a means of learning: An educational experiment in two high schools in Greece". Telematics and Informatics. 32: 118–128. doi:10.1016/j.tele.2014.05.001.

302. Pearce, Joshua M. 2012. "Building Research Equipment with Free, Open-Source Hardware." Science 337 (6100): 1303–1304

303. "Underwater City: 3D Printed Reef Restores Bahrain's Marine Life". ptc.com. 2013-08-01. Retrieved 2013-10-30.

304. Scopigno, R.; Cignoni, P.; Pietroni, N.; Callieri, M.; Dellepiane, M. (November 2015). "Digital Fabrication Techniques for Cultural Heritage: A Survey". Computer Graphics Forum: n/a. doi:10.1111/cgf.12781.

305. "Museum uses 3D printing to take fragile maquette by Thomas Hart Benton on tour through the States".

306. "British Museum releases 3D printer scans of artefacts". 2014-11-04.

307. "Threeding Uses Artec 3D Scanning Technology to Catalog 3D Models for Bulgaria's National Museum of Military History". 3dprint.com. 2015-02-20.

308. "$5,000 3D printer prints carbon-fiber parts". MarkForged.

309. "Researchers at UC Berkeley Create Bloom First Ever 3D-printed Cement Structure That Stands 9 Feet Tall". cbs sanfrancisco. 6 March 2015. Retrieved 23 April 2015.

310. Chino, Mike (9 March 2015). "UC Berkeley unveils 3D-printed "Bloom" building made of powdered cement". Retrieved 23 April 2015.

311. Fixsen, Anna (6 March 2015). "Print it Real Good: First Powder-Based 3D Printed Cement Structure Unveiled". Retrieved 23 April 2015.
312. "3D Printing Technology Insight Report, 2014, patent activity involving 3D-Printing from 1990–2013" (PDF). Retrieved 2014-06-10.
313. Thompson, Clive (30 May 2012). "3-D Printing's Legal Morass".
314. Weinberg, Michael (January 2013). "What's the Deal with copyright and 3D printing?" (PDF). Institute for Emerging Innovation. Retrieved 2013-10-30.
315. "Homeland Security bulletin warns 3D-printed guns may be 'impossible' to stop". Fox News. 2013-05-23. Retrieved 2013-10-30.
316. Cochrane, Peter (2013-05-21). "Peter Cochrane's Blog: Beyond 3D Printed Guns". TechRepublic. Retrieved 2013-10-30.
317. Gilani, Nadia (2013-05-06). "Gun factory fears as 3D blueprints put online by Defense Distributed". Metro.co.uk. Retrieved 2013-10-30.
318. "Liberator: First 3D-printed gun sparks gun control controversy". Digitaljournal.com. Retrieved 2013-10-30.
319. "First 3D Printed Gun 'The Liberator' Successfully Fired". International Business Times UK. 2013-05-07. Retrieved 2013-10-30.
320. "US demands removal of 3D printed gun blueprints". neurope.eu. Retrieved 2013-10-30.
321. "España y EE.UU. lideran las descargas de los planos de la pistola de impresión casera". ElPais.com. 2013-05-09. Retrieved 2013-10-30.
322. "Controlled by Guns". Quiet Babylon. 2013-05-07. Retrieved 2013-10-30.
323. "3dprinting". Joncamfield.com. Retrieved 2013-10-30.
324. "State Dept Censors 3D Gun Plans, Citing 'National Security'". News.antiwar.com. 2013-05-10. Retrieved 2013-10-30.
325. "Wishful Thinking Is Control Freaks' Last Defense Against 3D-Printed Guns". Reason.com. 2013-05-08. Retrieved 2013-10-30.
326. Lennard, Natasha (2013-05-10). "The Pirate Bay steps in to distribute 3-D gun designs". Salon.com. Archived from the original on 2013-05-19. Retrieved 2013-10-30.
327. "Sen. Leland Yee Proposes Regulating Guns From 3-D Printers". CBS Sacramento. 2013-05-08. Retrieved 2013-10-30.
328. "Schumer Announces Support For Measure To Make 3D Printed Guns Illegal".
329. "Four Horsemen of the 3D Printing Apocalypse". Makezine.com. 2011-06-30. Retrieved 2013-10-30.
330. Ball, James (10 May 2013). "US government attempts to stifle 3D-printer gun designs will ultimately fail". The Guardian. London.
331. Gadgets (2013-01-18). "Like It Or Not, 3D Printing Will Probably Be Legislated". TechCrunch. Retrieved 2013-10-30.
332. Klimas, Liz (2013-02-19). "Engineer: Don't Regulate 3D Printed Guns, Regulate

Explosive Gun Powder Instead". The Blaze. Retrieved 2013-10-30.

333. Beckhusen, Robert (2013-02-15). "3-D Printing Pioneer Wants Government to Restrict Gunpowder, Not Printable Guns". Wired. Retrieved 2013-10-30.
334. Bump, Philip (2013-05-10). "How Defense Distributed Already Upended the World". The Atlantic Wire. Archived from the original on 2013-05-19. Retrieved 2013-10-30.
335. "News". European Plastics News. Retrieved 2013-10-30.
336. "CDC - NIOSH Research Rounds - Volume 1, Issue 12, June 2016". www.cdc.gov. Retrieved 2016-11-16.
337. Albert 2011
338. "Confronting a New 'Era of Duplication'? 3D Printing, Replicating Technology and the Search for Authenticity in George O. Smith's Venus Equilateral Series". Durham University. Retrieved July 21, 2013.
339. "Materializing information: 3D printing and social change". Retrieved 2014-01-13.
340. "Additive Manufacturing: A supply chain wide response to economic uncertainty and environmental sustainability" (PDF). Retrieved 2014-01-11.
341. "Materializing information: 3D printing and social change". Retrieved 2014-03-30.
342. Kostakis, V. (2013): At the Turning Point of the Current Techno-Economic Paradigm: Commons-Based Peer Production, Desktop Manufacturing and the Role of Civil Society in the Perezian Framework.. In: TripleC, 11(1), 173–190.
343. Kostakis, V.; Papachristou, M. (2014). "Commons-based peer production and digital fabrication: The case of a RepRap-based, Lego-built 3D printing-milling machine". Telematics and Informatics. 31 (3): 434–443. doi:10.1016/j.tele.2013.09.006.
344. Kostakis, V; Fountouklis, M; Drechsler, W (2013). "Peer Production and Desktop Manufacturing: The Case of the Helix-T Wind Turbine Project". Science, Technology & Human Values. 38 (6): 773–800. doi:10.1177/0162243913493676.
345. Campbell, Thomas, Christopher Williams, Olga Ivanova, and Banning Garrett. (2011): Could 3D Printing Change the World? Technologies, Potential, and Implications of Additive Manufacturing ArchivedAugust 15, 2013, at the Wayback Machine.. Washington: Atlantic Council of the United States
346. Bradshaw, Simon, Adrian Bowyer, and Patrick Haufe (2010): The Intellectual Property Implications of Low-Cost 3D Printing. In: SCRIPTed 7
347. Gershenfeld, Neil (2007): FAB: The Coming Revolution on your Desktop: From Personal Computers to Personal Fabrication. Cambridge: Basic Books, p. 13–14
348. Larry Summers, The Inequality Puzzle, Democracy: A Journal of Ideas, Issue #32, Spring 2014
349. Michael Spence, Labor's Digital Displacement (2014-05-22), Project Syndicate
350. Sharma, Rakesh. "Can 3D Printing Reshape Manufacturing In America?". Forbes.com.
351. Nanomedicine, Volume I: Basic Capabilities, by Robert A. Freitas Jr. 1999,

ISBN 1-57059-645-X

352. Wagner V; Dullaart A; Bock AK; Zweck A. (2006). "The emerging nanomedicine landscape". Nat. Biotechnol. 24 (10): 1211–1217. doi:10.1038/nbt1006-1211. PMID 17033654.

353. Freitas RA Jr. (2005). "What is Nanomedicine?" (PDF). Nanomedicine: Nanotechnology, Biology and Medicine. 1 (1): 2–9. doi:10.1016/j.nano.2004.11.003. PMID 17292052.

354. Nanotechnology in Medicine and the Biosciences, by Coombs RRH, Robinson DW. 1996, ISBN 2-88449-080-9

355. "Market report on emerging nanotechnology now available". Market Report. US National Science Foundation. 25 February 2014. Retrieved 7 June 2016.

356. LaVan DA; McGuire T; Langer R. (2003). "Small-scale systems for in vivo drug delivery". Nat. Biotechnol. 21 (10): 1184–1191. doi:10.1038/nbt876. PMID 14520404.

357. Cavalcanti A; Shirinzadeh B; Freitas RA Jr.; Hogg T. (2008). "Nanorobot architecture for medical target identification". Nanotechnology. 19 (1): 015103(15pp). Bibcode:2008Nanot..19a5103C. doi:10.1088/0957-4484/19/01/015103.

358. Boisseau, P.; Loubaton, B. (2011). "Nanomedicine, nanotechnology in medicine". Comptes Rendus Physique. 12 (7): 620–636. doi:10.1016/j.crhy.2011.06.001.

359. Rao, Shasha; Tan, Angel; Thomas, Nicky; Prestidge, Clive. "Perspective and potential of oral lipid-based delivery to optimize pharmacological therapies against cardiovascular diseases". Journal of Controlled Release. 193: 174–187. doi:10.1016/j.jconrel.2014.05.013.

360. University of Waterloo, Nanotechnology in Targeted Cancer Therapy, https://www.youtube.com/watch?v=RBjWwlnq3cA 15 January 2010

361. Allen TM; Cullis PR (2004). "Drug Delivery Systems: Entering the Mainstream". Science. 303 (5665): 1818–1822. Bibcode:2004Sci...303.1818A. doi:10.1126/science.1095833. PMID 15031496.

362. Walsh MD; Hanna SK; Sen J; Rawal S; Cabral CB; Yurkovetskiy AV; Fram RJ; Lowinger TB; et al. (2012). "Pharmacokinetics and antitumor efficacy of XMT-1001, a novel, polymeric topoisomerase I inhibitor, in mice bearing HT-29 human colon carcinoma xenografts". Clin. Cancer Res. 18 (9): 2591–602. doi:10.1158/1078-0432.CCR-11-1554. PMID 22392910.

363. Chu KS; Hasan W; Rawal S; Walsh MD; Enlow EM; Luft JC; Bridges AS; Kuijer JL; et al. (Jul 2013). "Plasma, tumor and tissue pharmacokinetics of Docetaxel delivered via nanoparticles of different sizes and shapes in mice bearing SKOV-3 human ovarian carcinoma xenograft". Nanomedicine. 9 (5): 686–93. doi:10.1016/j.nano.2012.11.008. PMID 23219874.

364. Caron WP; Song G; Kumar P; Rawal S; Zamboni WC (2012). "Interpatient pharmacokinetic and pharmacodynamic variability of carrier-mediated anticancer agents". Clin. Pharmacol. Ther. 91 (5): 802–12. doi:10.1038/clpt.2012.12. PMID 22472987.

365. Bertrand N; Leroux JC (2011). "The journey of a drug carrier in the body: an anatomo-physiological perspective". Journal of Controlled Release. 161 (2): 152–63. doi:10.1016/j.jconrel.2011.09.098. PMID 22001607.

366. Nagy ZK, Zsombor K.; Balogh A; Vajna B; Farkas A; Patyi G; Kramarics A; Marosi G (2011). "Comparison of Electrospun and Extruded Soluplus-Based Solid Dosage Forms of Improved Dissolution". Journal of Pharmaceutical Sciences. 101 (1): n/a. doi:10.1002/jps.22731. PMID 21918982.

367. Minchin, Rod (2008). "Sizing up targets with nanoparticles". Nature Nanotechnology. 3 (1): 12–13. Bibcode:2008NatNa...3...12M. doi:10.1038/nnano.2007.433. PMID 18654442.

368. Banoee, M.; Seif, S.; Nazari, Z. E.; Jafari-Fesharaki, P.; Shahverdi, H. R.; Moballegh, A.; Moghaddam, K. M.; Shahverdi, A. R. (2010). "ZnO nanoparticles enhanced antibacterial activity of ciprofloxacin against Staphylococcus aureus and Escherichia coli". J Biomed Mater Res B Appl Biomater. 93 (2): 557–61. doi:10.1002/jbm.b.31615. PMID 20225250.

369. Seil JT; Webster TJ (2012). "Antimicrobial applications of nanotechnology: methods and literature". Int J Nanomedicine. 7: 2767–81. doi:10.2147/IJN.S24805. PMID 22745541.

370. Borzabadi-Farahani A; Borzabadi E; Lynch E. (2013). "Nanoparticles in orthodontics, a review of antimicrobial and anti-caries applications". Acta Odontol Scand. 72 (6): 413–417. doi:10.3109/00016357.2013.859728. PMID 24325608.

371. Conde, J.; de la Fuente, JM; Baptista, PV. "Nanomaterials for reversion of multidrug resistance in cancer: a new hope for an old idea?." Front. Pharmacol.2013. Vol 4 No 134.

372. Sanjeev Soni, Himanshu Tyagi, Robert A. Taylor, Amod Kumar, Role of optical coefficients and healthy tissue-sparing characteristics in gold nanorod-assisted thermal therapy, International Journal of Hyperthermia, 2013, Vol. 29, No. 1 , Pages 87-97

373. Nanospectra Biosciences, Inc. – Publications (http://www.nanospectra.com/clinicians/spublications.html)

374. Mozafari, M.R. (ed), (2006) Nanocarrier Technologies: Frontiers of Nanotherapy (Chapters 1 and 2) pages 10–11, 25–34

375. Bertrand N; Bouvet C; Moreau P & Leroux JC (2010). "Transmembrane pH-Gradient Liposomes To Treat Cardiovascular Drug Intoxication". ACS Nano. 4 (12): 7552–7558. doi:10.1021/nn101924a. PMID 21067150.

376. Mashaghi S.; Jadidi T.; Koenderink G.; Mashaghi A. (2013). "Lipid Nanotechnology". Int. J. Mol. Sci. 2013 (14): 4242–4282. doi:10.3390/ijms14024242. PMC 3588097 . PMID 23429269.

377. Conde, J; Oliva, N; Atilano, M; Song, HS; Artzi, N (2015). "Self-assembled RNA-triple-helix hydrogel scaffold for microRNA modulation in the tumour microenvironment". Nat Mater. 15: 353–63. doi:10.1038/nmat4497. PMID 26641016.

378. FDA (October 2012). “Highlights of Prescribing Information, Abraxane for Injectable Suspension” (PDF).

379. “Paclitaxel (Abraxane)”. U.S. Food and Drug Administration. 11 October 2012. Retrieved 10 December 2012.

380. FDA Press Announcements (6 September 2013). “FDA approves Abraxane for late-stage pancreatic cancer”. FDA.

381. Martis, Elvis A.; Badve, Rewa R.; Degwekar, Mukta D. (January 2012). “Nanotechnology based devices and applications in medicine: An overview”. Chronicles of Young Scientists. 3 (1): 68–73. doi:10.4103/2229-5186.94320.

382. News Release (22 October 2015). “FDA approves new treatment for advanced pancreatic cancer”. FDA.

383. Gibney, Michael (18 April 2013). “Cornell nanosized ‘dots’ for diagnostics may now deliver drugs”. fiercedrugdelivery.com. Retrieved 17 June 2013.

384. Elvidge, Suzanne (11 November 2012). “Bacterial ‘minicells’ deliver cancer drugs straight to the target”. fiercedrugdelivery.com. Retrieved 10 December 2012.

385. “First trial in humans of ‘minicells’: a completely new way of delivering anti-cancer drugs”. fiercedrugdelivery.com. 12 November 2012. Retrieved 10 December 2012.

386. Press Release (19 June 2014). “Kala Pharmaceuticals Initiates Phase3 Clinical Trial for Treatment of Post-Surgical Ocular Inflammation and Phase 2 Clinical Trial in Dry Eye Disease” (PDF). Kala Pharmaceuticals.

387. Press Release (1 April 2015). “Kala Pharmaceuticals Announces Positive Results from Phase 3 Trial of KPI-121 in Cataract Surgery”(PDF). Kala Pharmaceuticals.

388. Wang AZ, Langer R, Farokhzad OC (2012). “Nanoparticle delivery of cancer drugs”. Annu. Rev. Med. 63: 185–98. doi:10.1146/annurev-med-040210-162544. PMID 21888516.

389. Pérez-Herrero E, Fernández-Medarde A (2015). “Advanced targeted therapies in cancer: Drug nanocarriers, the future of chemotherapy”. Eur J Pharm Biopharm. 93: 52–79. doi:10.1016/j.ejpb.2015.03.018. PMID 25813885.

390. Nguyen-Ngoc T, Raymond E (2015). “Reinvention of chemotherapy: drug conjugates and nanoparticles”. Curr Opin Oncol. 27: 232–42. doi:10.1097/CCO.0000000000000183. PMID 25783982.

391. Ghaz-Jahanian MA, Abbaspour-Aghdam F, Anarjan N, Berenjian A, Jafarizadeh-Malmiri H (2015). “Application of chitosan-based nanocarriers in tumor-targeted drug delivery”. Mol. Biotechnol. 57 (3): 201–18. doi:10.1007/s12033-014-9816-3. PMID 25385004.

392. Nie, Shuming; Yun Xing; Gloria J. Kim & Jonathan W. Simmons (2007). “Nanotechnology Applications in Cancer”. Annual Review of Biomedical Engineering. 9 (1): 257–88. doi:10.1146/annurev.bioeng.9.060906.152025. PMID 17439359.

393. Chidambaram, M.; Manavalan, R.; Kathiresan, K. (2011). “Nanotherapeutics to overcome conventional cancer chemotherapy limitations”. Journal of pharmacy &

pharmaceutical sciences : a publication of the Canadian Society for Pharmaceutical Sciences, Societe canadienne des sciences pharmaceutiques. 14 (1): 67–77. PMID 21501554.

394. Loo C; Lin A; Hirsch L; Lee MH; Barton J; Halas N; West J; Drezek R. (2004). "Nanoshell-enabled photonics-based imaging and therapy of cancer". Technol Cancer Res Treat. 3 (1): 33–40. doi:10.1177/153303460400300104. PMID 14750891.

395. Conde J, Tian F, Hernández Y, Bao C, Cui D, Janssen KP, Ibarra MR, Baptista PV, Stoeger T, de la Fuente JM. In vivo tumor targeting via nanoparticle-mediated therapeutic siRNA coupled to inflammatory response in lung cancer mouse models. Biomaterials. 2013;34(31):7744-53. doi:10.1016/j.biomaterials.2013.06.041

396. Yasitha L Hewakuruppu et al., Plasmonic " pump – probe " method to study semi-transparent nanofluids, Applied Optics, 52(24) 6041-6050.

397. Coffey, Rebecca (August 2010). "20 Things You Didn't Know About Nanotechnology". Discover. 31 (6): 96.

398. Zheng G; Patolsky F; Cui Y; Wang WU; Lieber CM (2005). "Multiplexed electrical detection of cancer markers with nanowire sensor arrays". Nat. Biotechnol. 23 (10): 1294–1301. doi:10.1038/nbt1138. PMID 16170313.

399. Hall, J. Storrs (2005). Nanofuture : what's next for nanotechnology. Amherst, NY: Prometheus Books. ISBN 978-1591022879.

400. "Drug Store Cancer Tests". Technology Review. 2005-10-31. Retrieved 2009-10-08.

401. Keller, John (2013). "Nanotechnology has also helped to personalize oncology for the detection, diagnosis, and treatment of cancer. It is now able to be tailored to each individual's tumor for better performance". Military & Aerospace Electronics. 23 (6): 27.

402. Herrmann I. K.; et al. (2013). "Nanomagnet-based removal of lead and digoxin from living rats". Nanoscale. 5: 8718–8723. doi:10.1039/c3nr02468g.

403. Kang J. H.; et al. (2014). "An extracorporeal blood-cleansing device for sepsis therapy". Nat. Med. 20: 1211–1216. doi:10.1038/nm.3640.

404. Berry C. C.; Curtis A. S. G. (2003). "Functionalisation of magnetic nanoparticles for applications in biomedicine". J. Phys. Appl. Phys. 36: R198. doi:10.1088/0022-3727/36/13/203.

405. Herrmann I. K.; et al. (2013). "Endotoxin Removal by Magnetic Separation-Based Blood Purification". Adv. Healthcare Mater. 2: 829–835. doi:10.1002/adhm.201200358.

406. Lee J.-J.; et al. (2014). "Synthetic Ligand-Coated Magnetic Nanoparticles for Microfluidic Bacterial Separation from Blood". Nano Lett. 14: 1–5. doi:10.1021/nl3047305.

407. Schumacher C. M.; et al. (2013). "Quantitative Recovery of Magnetic Nanoparticles from Flowing Blood: Trace Analysis and the Role of Magnetization". Adv. Funct. Mater. 23: 4888–4896. doi:10.1002/adfm.201300696.

408. Yung C. W.; Fiering J.; Mueller A. J.; Ingber D. E. (2009). "Micromagnetic–microfluidic blood cleansing device". Lab. Chip. 9: 1171–7. doi:10.1039/b816986a. PMID 19370233.

409. Herrmann, I. K., Grass, R. N. & Stark, W. J. High-strength metal nanomagnets for diagnostics and medicine: carbon shells allow long-term stability and reliable linker chemistry. Nanomed. 4, 787–798 (2009).

410. Stacy Shepherd,"Harvard Engineers Invented an Artificial Spleen to Treat Sepsis", Boston Magazine. Retrieved on 20 April 2015.

411. Lalwani Gaurav; Henslee Allan M.; Farshid Behzad; Lin Liangjun; Kasper F. Kurtis, Yi-; Qin Xian; Mikos Antonios G.; Sitharaman Balaji; et al. (2013). "Two-dimensional nanostructure-reinforced biodegradable polymeric nanocomposites for bone tissue engineering". Biomacromolecules. 14 (3): 900–909. doi:10.1021/bm301995s. PMC 3601907 . PMID 23405887.

412. Lalwani, Gaurav (September 2013). "Tungsten disulfide nanotubes reinforced biodegradable polymers for bone tissue engineering". Acta Biomaterialia. 9 (9): 8365–8373. doi:10.1016/j.actbio.2013.05.018. PMC 3732565 . PMID 23727293.

413. Gobin AM; O'Neal DP; Watkins DM; Halas NJ; Drezek RA; West JL (2005). "Near infrared laser-tissue welding using nanoshells as an exogenous absorber". Lasers Surg Med. 37 (2): 123–9. doi:10.1002/lsm.20206. PMID 16047329.

414. "A nanoscale biofuel cell for self-powered nanotechnology devices".

415. Nanomedicine, Volume IIA: Biocompatibility, by Robert A. Freitas Jr. 2003, ISBN 1-57059-700-6

416. Freitas, Robert A., Jr. (2005). "Current Status of Nanomedicine and Medical Nanorobotics" (PDF). Journal of Computational and Theoretical Nanoscience. 2 (4): 1–25. doi:10.1166/jctn.2005.001.

417. "Nanofactory Collaboration".

418. Kurzweil, Ray (2005). The Singularity Is Near. New York City: Viking Press. ISBN 978-0-670-03384-3. OCLC 57201348.

419. Richard P. Feynman (December 1959). "There's Plenty of Room at the Bottom". Retrieved March 2010. Check date values in: |access-date= (help)

CHAPTER 9

Cloud Robotics

Chapter Contents

As cloud computing captures the public imagination, it also has become the latest area of serious study in robotics on numerous college campuses, including Philadelphia's Drexel University and the University of Minnesota.

INTRODUCTION

Cloud robotics is a field of robotics that attempts to invoke cloud technologies such as cloud computing, cloud storage, and other Internet technologies centred on the benefits of converged infrastructure and shared services for robotics. When connected to the cloud, robots can benefit from the powerful computational, storage, and communications resources of modern data centre in the cloud, which can process and share information from various robots or agent (other machines, smart objects, humans, etc.). Humans can also delegate tasks to robots remotely through networks. Cloud computing technologies enable robot systems to be endowed with powerful capability whilst reducing costs through cloud technologies. Thus, it is possible to build lightweight, low cost, smarter robots have intelligent "brain" in the cloud. The "brain" consists of data center, knowledge base, task planners, deep learning, information processing, environment models, communication support etc.

Components

A cloud for robots potentially has at least six significant components:

- Offering a global library of images, maps, and object data, often with geometry and mechanical properties, expert system, knowledge base (i.e. semantic web, data centres)
- Massively-parallel computation on demand for sample-based statistical modelling and motion planning, task planning, multi-robot collaboration, scheduling and coordination of system
- Robot sharing of outcomes, trajectories, and dynamic control policies and robot learning support
- Human sharing of "open-source" code, data, and designs for programming, experimentation, and hardware construction
- On-demand human guidance and assistance for evaluation, learning, and error recovery;
- Augmented human–robot interaction through various way (Semantics knowledge base, Apple SIRI like service etc.).

HUMAN–ROBOT INTERACTION

Human–robot interaction is the study of interactions between humans and robots. It is often referred as HRI by researchers. Human–robot interaction is a multidisciplinary field with contributions from human–computer interaction, artificial intelligence, robotics, natural language understanding, design, and social sciences.

Origins

Human–robot interaction has been a topic of both science fiction and academic speculation even before any robots existed. Because HRI depends on a knowledge of (sometimes natural) human communication, many aspects of HRI are continuations of human communications topics that are much older than robotics per se.

The origin of HRI as a discrete problem was stated by 20th-century author Isaac Asimov in 1941, in his novel *I, Robot.* He states the Three Laws of Robotics as,

- A robot may not injure a human being or, through inaction, allow a human being to come to harm.
- A robot must obey any orders given to it by human beings, except where such orders would conflict with the First Law.
- A robot must protect its own existence as long as such protection does not conflict with the First or Second Law.

These three laws of robotics determine the idea of safe interaction. The closer the human and the robot get and the more intricate the relationship becomes, the more the risk of a human being injured rises. Nowadays in advanced societies, manufacturers employing robots solve this issue by not letting humans and robots share the workspace at any time. This is achieved by defining safe zones using lidar sensors or physical cages. Thus the presence of humans is completely forbidden in the robot workspace while it is working.

With the advances of artificial intelligence, the autonomous robots could eventually have more proactive behaviors, planning their motion in complex unknown environments. These new capabilities keep safety as the primary issue and efficiency as secondary. To allow this new generation of robot, research is being conducted on human detection, motion planning, scene reconstruction, intelligent behavior through task planning and compliant behavior using force control (impedance or admittance control schemes).

The goal of HRI research is to define models of humans' expectations regarding robot interaction to guide robot design and algorithmic development that would allow more natural and effective interaction between humans and robots. Research ranges from how humans work with remote, tele-operated unmanned vehicles to peer-to-peer collaboration with anthropomorphic robots.

Many in the field of HRI study how humans collaborate and interact and use those studies to motivate how robots should interact with humans.

The Goal of Friendly Human–Robot Interactions

Kismet can produce a range of facial expressions.

Robots are artificial agents with capacities of perception and action in the physical world often referred by researchers as workspace. Their use has been generalized in factories but nowadays they tend to be found in the most technologically advanced societies in such critical domains as search and rescue, military battle, mine and bomb detection, scientific exploration, law enforcement, entertainment and hospital care. These new domains of applications imply a closer interaction with the user. The concept of closeness is to be taken in its full meaning, robots and humans share the workspace but also share goals in terms of task achievement. This close interaction needs new theoretical models, on one hand for the robotics scientists who work to improve the robots utility and on the other hand to evaluate the risks and benefits of this new "friend" for our modern society.

With the advance in AI, the research is focusing on one part towards the safest physical interaction but also on a socially correct interaction, dependent on cultural criteria. The goal is to build an intuitive, and easy communication with the robot through speech, gestures, and facial expressions.

Dautenhahn refers to friendly Human–robot interaction as "Robotiquette" defining it as the "social rules for robot behaviour (a 'robotiquette') that is comfortable and acceptable to humans" The robot has to adapt itself to our way of expressing desires and orders and not the contrary. But every day environments such as homes have much more complex social rules than those implied by factories or even military environments. Thus, the robot needs perceiving and understanding capacities to build dynamic models of its surroundings. It needs to categorize objects, recognize and locate humans and further their emotions. The need for dynamic capacities pushes forward every sub-field of robotics.

On the other end of HRI research the cognitive modelling of the "relationship" between human and the robots benefits the psychologists and robotic researchers the user study are often of interests on both sides. This research endeavours part of human society.

General HRI Research

HRI research spans a wide range of field, some general to the nature of HRI.

Methods for Perceiving Humans

Most methods intend to build a 3D model through vision of the environment. The proprioception sensors permit the robot to have information over its own state. This information is relative to a reference.

Methods for perceiving humans in the environment are based on sensor information. Research on sensing components and software led by Microsoft provide useful results for extracting the human kinematics (see Kinect). An example of older technique is to use colour information for example the fact that for light skinned people the hands are lighter than the clothes worn. In any case a human modelled a priori can then be fitted to the sensor data. The robot builds or has (depending on the level of autonomy the robot has) a 3D mapping of its surroundings to which is assigned the humans locations.

A speech recognition system is used to interpret human desires or commands. By combining the information inferred by proprioception, sensor and speech the human position and state (standing, seated).

Methods for Motion Planning

Motion planning in dynamic environment is a challenge that is for the moment only achieved for 3 to 10 degrees of freedom robots. Humanoid robots or even 2 armed robots that can have up to 40 degrees of freedom are unsuited for dynamic environments with today's technology. However lower-dimensional robots can use potential field method to compute trajectories avoiding collisions with human.

Cognitive Models and Theory of Mind

Humans exhibit negative social and emotional responses as well as decreased trust toward some robots that closely, but imperfectly, resemble humans; this phenomenon has been termed the "Uncanny Valley."

A lot of data has been gathered with regards to user studies. For example, when users encounter proactive behaviour on the part of the robot and the robot does not respect a safety distance, penetrating the user space, he or she might express fear. This is dependent on one person to another. Only intensive experiment can permit a more precise model. It has been shown that when a robot has no particular use, negative feelings are often expressed. The robot is perceived as useless and its presence becomes annoying. In another experiment, it has occurred that people tend to attribute to the robot personality characteristics that were not implemented.

Application-Oriented HRI Research

In addition to general HRI research, researchers are currently exploring application areas for human-robot interaction systems. Application-oriented research is used to help bring current robotics technologies to bear against problems that exist in today's society. While human-robot interaction is still a rather young area of interest, there is active development and research in many areas.

HRI/OS Research

The Human-Robot Interaction Operating System(HRI/OS), "provides a structured software framework for building human-robot teams, supports a variety of user interfaces, enables humans and robots to engage in task-oriented dialogue, and facilitates integration of robots through an extensible API".

Search and Rescue

First responders face great risks in search and rescue (SAR) settings, which typically involve environments that are unsafe for a human to travel. In addition, technology offers tools for observation that can greatly speed-up and improve the accuracy of human perception. Robots can be used to address these concerns . Research in this area includes efforts to address robot sensing, mobility, navigation, planning, integration, and tele-operated control.

SAR robots have already been deployed to environments such as the Collapse of the World Trade Center.

Other application areas include:

- Entertainment
- Education
- Field robotics
- Home and companion robotics
- Hospitality
- Rehabilitation and Elder Care
- Robot Assisted Therapy (RAT)

Applications

Autonomous mobile robots: Google's self-driving cars are cloud robots. The cars use the network to access Google's enormous database of maps and satellite and environment model (like Streetview) and combines it with streaming data from GPS, cameras, and 3D sensors to monitor its own position within centimetres, and with past and current traffic patterns to avoid collisions. Each car can learn something about environments, roads, or driving, or conditions, and it sends the information to the Google cloud, where it can be used to improve the performance of other cars.

Cloud medical robots: a medical cloud (also called a healthcare cluster) consists of various services such as a disease archive, electronic medical records, a patient health management system, practice services, analytics services, clinic solutions, expert systems, etc. A robot can connect to the cloud to provide clinical service to patients, as well as deliver assistance to doctors (e.g. a co-surgery robot). Moreover, it also provides a collaboration service by sharing information between doctors and care givers about clinical treatment. Assistive robots: A domestic robot can be employed for healthcare and life monitoring for elderly people. The system collects the health status of users and exchange information with cloud expert system or doctors to facilitate elderly peoples life, especially for those with chronic diseases. For example, the robots are able to provide support to prevent the elderly from falling down, emergency healthy support such as heart disease, blooding disease. Care givers of elderly people can also get notification when in emergency from the robot through network.

Industrial robots: As highlighted by the Germany Industry 4.0 Plan "Industry is on the threshold of the fourth industrial revolution. Driven by the Internet, the real and virtual worlds are growing closer and closer together to form the Internet of Things. Industrial production of the future will be characterised by the strong individualisation of products under the conditions of highly flexible (large series) production, the extensive integration of customers and business partners in business and value-added processes, and the linking of production and high-quality services leading to so-called hybrid products." In manufacturing, such cloud based robot systems could learn to handle tasks such as threading wires or cables, or aligning gaskets from professional knowledge base. A group of robots can share information for some collaborative tasks. Even more, a consumer is able to order customised product to manufacturing robots directly with online order system. Another potential paradigm is shopping-delivery robot system- once an order is placed, a warehouse robot dispatches the item to an autonomous car or autonomous drone to delivery it to its recipient.

Research

RoboEarth was funded by the European Union's Seventh Framework Programme for research, technological development projects, specifically to explore the field of cloud robotics. The goal of RoboEarth is to allow robotic systems to benefit from the experience of other robots, paving the way for rapid advances in machine cognition and behaviour, and ultimately, for more subtle and sophisticated human-machine interaction. RoboEarth offers a Cloud Robotics infrastructure. RoboEarth's World-Wide-Web style database stores knowledge generated by humans – and robots – in a machine-readable format. Data stored in the RoboEarth knowledge base include software components, maps for navigation (e.g., object locations, world models), task knowledge (e.g., action recipes, manipulation strategies), and object recognition models (e.g., images, object models). The RoboEarth Cloud Engine includes support for mobile robots, autonomous vehicles, and drones, which require lots of computation for navigation.

Rapyuta is an open source cloud robotics framework based on RoboEarth Engine developed by the robotics researcher at ETHZ. Within the framework, each robot connected to Rapyuta can have a secured computing environment (rectangular boxes) giving them the ability to move their heavy computation into the cloud. In addition, the computing environments are tightly interconnected with each other and have a high bandwidth connection to the RoboEarth knowledge repository.

KnowRob is an extensional project of RoboEarth. It is a knowledge processing system that combines knowledge representation and reasoning methods with techniques for acquiring knowledge and for grounding the knowledge in a physical system and can serve as a common semantic framework for integrating information from different sources.

RoboBrain is a large-scale computational system that learns from publicly available Internet resources, computer simulations, and real-life robot trials. It accumulates everything robotics into a comprehensive and interconnected knowledge base. Applications include prototyping for robotics research, household robots, and self-driving cars. The goal is as direct as the project's name—to create a centralised, always-online brain for robots to tap into. The project is dominated by Stanford University and Cornel University. And the project is supported by the National Science Foundation, the Office of Naval Research, the Army Research Office, Google, Microsoft, Qualcomm, the Alfred P. Sloan Foundation and the National Robotics Initiative, whose goal is to advance robotics to help make the United States more competitive in the world economy.

MyRobots is a service for connecting robots and intelligent devices to the Internet. It can be regarded as a social network for robots and smart objects (i.e. Facebook for robots). With socialising, collaborating and sharing, robots can benefit from those interactions too by sharing their sensor information giving insight on their perspective of their current state.

COALAS is funded by the INTERREG IVA France (Channel) – England European cross-border co-operation programme. The project aims to develop new technologies for handicapped people through social and technological innovation and through the users' social and psychological integrity. Objectives is to produce a cognitive ambient assistive living system with Healthcare cluster in cloud with domestic service robots like humanoid, intelligent wheelchair which connect with the cloud.

ROS(Robot Operating System) provides an eco-system to support cloud robotics. ROS is a flexible and distributed framework for robot software development. It is a collection of tools, libraries, and conventions that aim to simplify the task of creating complex and robust robot behaviour across a wide variety of robotic platforms. A library for ROS that is a pure Java implementation, called rosjava, allows Android applications to be developed for robots. Since Android has a booming market and billion users, it would be significant in the field of Cloud Robotics.

Limitations of Cloud Robotics

Though robots can benefit from various advantages of cloud computing, cloud is not the solution to all of robotics.

- Controlling a robot's motion which relies heavily on sensors and feedback of controller won't benefit much from the cloud.
- Cloud-based applications can get slow or unavailable due to high-latency responses or network hitch. If a robot relies too much on the cloud, a fault in the network could leave it "brainless."
- Tasks that involve real-time execution require on-board processing.

Challenges

The research and development of cloud robotics has following potential issues and challenges:

- Scalable parallelisation-grid-based computing, parallelisation schemes scale with the size of automation infrastructure.
- Effective load balancing: Balancing operations between local and cloud computation.
- Knowledge bases and representations
- Collective learning for automation in cloud
- Infrastructure/Platform or Software as a Service
- Internet of Things for robotics
- Integrated and collaborative fault tolerant control
- Big Data: Data, collected and/or disseminated over large, accessible networks can enable decisions for classification problems or reveal patterns.
- Wireless communication, Connectivity to the cloud
- System architectures of robot cloud
- Open-source, open-access infrastructures

- Workload Sharing
- Standards and Protocol

Risks

Environmental security - The concentration of computing resources and users in a cloud computing environment also represents a concentration of security threats. Because of their size and significance, cloud environments are often targeted by virtual machines and bot malware, brute force attacks, and other attacks.

Data privacy and security - Hosting confidential data with cloud service providers involves the transfer of a considerable amount of an organisation's control over data security to the provider. For example, every cloud contains a huge information from the clients include personal data. If a household robot is hacked, users could have risk of their personal privacy and security, like house layout, life snapshot, home-view, etc. It may be accessed and leaked to the world around by criminals. Another problems is once a robot is hacked and controlled by someone else, which may put the user in danger.

Ethical problems - Some ethics of robotics, especially for cloud based robotics must be considered. Since a robot is connected via networks, it has risk to be accessed by other people. If a robot is out of control and carries out illegal activities, who should be responsible for it.

Special Issue on Cloud Robotics and Automation- A special issue of the IEEE Transactions on Automation Science and Engineering, April 2015.

Cloud Robotics-Enable cloud computing for robots. The author proposed some paradigms of using cloud computing in robotics. Some potential field and challenges were coined. R. Li 2014.

2013 IEEE IROS Workshop on Cloud Robotics. Tokyo. November 2013.

NRI Workshop on Cloud Robotics: Challenges and Opportunities- February 2013.

Robot APP Store Robot Applications in Cloud, provide applications for robot just like computer/phone app.

DARPA Cloud Robotics.

A Roadmap for U.S. Robotics From Internet to Robotics 2013 Edition- by Georgia Institute of Technology, Carnegie Mellon University Robotics Technology Consortium, University of Pennsylvania, University of Southern California, Stanford University, University of California–Berkeley, University of Washington, Massachusetts Institute of TechnologyUS and Robotics OA US. The Roadmap highlighted "Cloud" Robotics and Automation for Manufacturing in the future years.

Cloud-Based Robot Grasping with the Google Object Recognition Engine.

National Robotics Initiative of US announced in 2011 aimed to explore how robots can enhance the work of humans rather than replacing them. It claims that next generation of robots are more aware than oblivious, more social than solitary.

James J. Kuffner, a former CMU robotics professor, and now research scientist at Google, spoke on cloud robotics in IEEE/RAS International Conference on Humanoid Robotics 2010. It describes "a new approach to robotics that takes advantage of the Internet as a resource for massively parallel computation and sharing of vast data resources."

Ryan Hickman, a Google Product Manager, led an internal volunteer effort in 2010 to connect robots with the Google's cloud services.This work was later expanded to include open source ROS support and was demonstrated on stage by Ryan Hickman, Damon Kohler, Brian Gerkey, and Ken Conley at Google I/O 2011.

The IEEE RAS Technical Committee on Internet and Online Robots was founded by Ken Goldberg and Roland Siegwart et al. in May 2001. The committee then expanded to IEEE Society of Robotics and Automation's Technical Committee on Networked Robots in 2004.

OUTLINE OF ROBOTICS

Robotics is a branch of mechanical engineering, electrical engineering and computer science that deals with the design, construction, operation, and application of robots, as well as computer systems for their control, sensory feedback, and information processing. These technologies deal with automated machines that can take the place of humans in dangerous environments or manufacturing processes, or resemble humans in appearance, behaviour, and or cognition. Many of today's robots are inspired by nature contributing to the field of bio-inspired robotics.

The word "robot" was introduced to the public by Czech writer Karel Čapek in his play R.U.R. (Rossum's Universal Robots), published in 1920. The term «robotics» was coined by Isaac Asimov in his 1941 science fiction short-story «Liar!»

Nature of Robotics

Robotics can be described as:

- An applied science – scientific knowledge transferred into a physical environment.
    - A branch of computer science –
    - A branch of electrical engineering –
    - A branch of mechanical engineering –
- Research and development –
- A branch of technology –

Branches of Robotics

- Adaptive control – control method used by a controller which must adapt to a controlled system with parameters which vary, or are initially uncertain. For example, as an aircraft flies, its mass will slowly decrease as a result of fuel consumption; a control law is needed that adapts itself to such changing conditions.

- Aerial robotics – development of unmanned aerial vehicles (UAVs), commonly known as drones, aircraft without a human pilot aboard. Their flight is controlled either autonomously by onboard computers or by the remote control of a pilot on the ground or in another vehicle.
- Android science – interdisciplinary framework for studying human interaction and cognition based on the premise that a very humanlike robot (that is, an android) can elicit human-directed social responses in human beings.
- Anthrobotics – science of developing and studying robots that are either entirely or in some way human-like.
- Artificial intelligence – the intelligence of machines and the branch of computer science that aims to create it.
- Autonomous car – an autonomous vehicle capable of fulfilling the human transportation capabilities of a traditional car
- Autonomous research robotics –
- Bayesian network –
- BEAM robotics – a style of robotics that primarily uses simple analogue circuits instead of a microprocessor in order to produce an unusually simple design (in comparison to traditional mobile robots) that trades flexibility for robustness and efficiency in performing the task for which it was designed.
- Behavior-based robotics – the branch of robotics that incorporates modular or behavior based AI (BBAI).
- Bio-inspired robotics – making robots that are inspired by biological systems. Biomimicry and bio-inspired design are sometimes confused. Biomimicry is copying the nature while bio-inspired design is learning from nature and making a mechanism that is simpler and more effective than the system observed in nature.
- Biomimetic – see Bionics.
- Biomorphic robotics – a sub-discipline of robotics focused upon emulating the mechanics, sensor systems, computing structures and methodologies used by animals.
- Bionics – also known as biomimetics, biognosis, biomimicry, or bionical creativity engineering is the application of biological methods and systems found in nature to the study and design of engineering systems and modern technology.
- Biorobotics – a study of how to make robots that emulate or simulate living biological organisms mechanically or even chemically.
- Cloud robotics – is a field of robotics that attempts to invoke cloud technologies such as cloud computing, cloud storage, and other Internet technologies centered around the benefits of converged infrastructure and shared services for robotics.
- Cognitive robotics – views animal cognition as a starting point for the development of robotic information processing, as opposed to more traditional Artificial

Intelligence techniques.

- Clustering –
- Computational neuroscience – study of brain function in terms of the information processing properties of the structures that make up the nervous system.
- Robot control – a study of controlling robots
- Robotics conventions –
- Data mining Techniques –
- Degrees of freedom – in mechanics, the degree of freedom (DOF) of a mechanical system is the number of independent parameters that define its configuration. It is the number of parameters that determine the state of a physical system and is important to the analysis of systems of bodies in mechanical engineering, aeronautical engineering, robotics, and structural engineering.
- Developmental robotics – a methodology that uses metaphors from neural development and developmental psychology to develop the mind for autonomous robots
- Digital control – a branch of control theory that uses digital computers to act as system controllers.
- Digital image processing – the use of computer algorithms to perform image processing on digital images.
- Dimensionality reduction – the process of reducing the number of random variables under consideration, and can be divided into feature selection and feature extraction.
- Distributed robotics –
- Electronic stability control – is a computerized technology that improves the safety of a vehicle's stability by detecting and reducing loss of traction (skidding).
- Evolutionary computation –
- Evolutionary robotics – a methodology that uses evolutionary computation to develop controllers for autonomous robots
- Extended Kalman filter –
- Flexible Distribution functions –
- Feedback control and regulation –
- Human–computer interaction – a study, planning and design of the interaction between people (users) and computers
- Human robot interaction – a study of interactions between humans and robots
- Kinematics – study of motion, as applied to robots. This includes both the design of linkages to perform motion, their power, control and stability; also their planning, such as choosing a sequence of movements to achieve a broader task.
- Laboratory robotics – the act of using robots in biology or chemistry labs
- Robot learning – learning to perform tasks such as obstacle avoidance, control and various other motion-related tasks

- Manifold learning –
- Direct manipulation interface – In computer science, direct manipulation is a human–computer interaction style which involves continuous representation of objects of interest and rapid, reversible, and incremental actions and feedback. The intention is to allow a user to directly manipulate objects presented to them, using actions that correspond at least loosely to the physical world.
- Robotic mapping – the goal for an autonomous robot to be able to construct (or use) a map or floor plan and to localize itself in it
- Microrobotics – a field of miniature robotics, in particular mobile robots with characteristic dimensions less than 1 mm
- Motion planning – (a.k.a., the "navigation problem", the "piano mover's problem") is a term used in robotics for the process of detailing a task into discrete motions.
- Motor control – information processing related activities carried out by the central nervous system that organize the musculoskeletal system to create coordinated movements and skilled actions.
- Nanorobotics – the emerging technology field creating machines or robots whose components are at or close to the scale of a nanometer (10−9 meters).
- Artificial neural networks – a mathematical model inspired by biological neural networks.
- Passive dynamics – refers to the dynamical behavior of actuators, robots, or organisms when not drawing energy from a supply (e.g., batteries, fuel, ATP).
- Reinforcement learning – an area of machine learning in computer science, concerned with how an agent ought to take actions in an environment so as to maximize some notion of cumulative reward.
- Programming by Demonstration – an End-user development technique for teaching a computer or a robot new behaviors by demonstrating the task to transfer directly instead of programming it through machine commands.
- Robot kinematics – applies geometry to the study of the movement of multi-degree of freedom kinematic chains that form the structure of robotic systems.
- Robot locomotion – collective name for the various methods that robots use to transport themselves from place to place.
- Rapid prototyping – automatic construction of physical objects via additive manufacturing from virtual models in computer aided design (CAD) software, transforming them into thin, virtual, horizontal cross-sections and then producing successive layers until the items are complete. As of June 2011, used for making models, prototype parts, and production-quality parts in relatively small numbers.
- Robot programming –
- Sensors – (also called detector) is a converter that measures a physical quantity and converts it into a signal which can be read by an observer or by an (today mostly electronic) instrument.
- Simultaneous localization and mapping – a technique used by robots and autonomous vehicles to build up a map within an unknown environment (without

a priori knowledge), or to update a map within a known environment (with a priori knowledge from a given map), while at the same time keeping track of their current location.

- Software engineering – the application of a systematic, disciplined, quantifiable approach to the design, development, operation, and maintenance of software, and the study of these approaches; that is, the application of engineering to software.
- Speech processing – study of speech signals and the processing methods of these signals. The signals are usually processed in a digital representation, so speech processing can be regarded as a special case of digital signal processing, applied to speech signal. Aspects of speech processing includes the acquisition, manipulation, storage, transfer and output of digital speech signals.
- Support vector machines – supervised learning models with associated learning algorithms that analyze data and recognize patterns, used for classification and regression analysis.
- Robotic surgery – computer-assisted surgery, and robotically-assisted surgery are terms for technological developments that use robotic systems to aid in surgical procedures.
    - Remote surgery – (also known as telesurgery) is the ability for a doctor to perform surgery on a patient even though they are not physically in the same location.
    - Robot-assisted heart surgery –
- Swarm robotics – involves large numbers of mostly simple physical robots. Their actions may seek to incorporate emergent behavior observed in social insects (swarm intelligence).
    - Ant robotics – swarm robots that can communicate via markings, similar to ants that lay and follow pheromone trails.
- Telepresence – refers to a set of technologies which allow a person to feel as if they were present, to give the appearance of being present, or to have an effect, via telerobotics, at a place other than their true location.
- Intelligent vehicle technologies – comprise electronic, electromechanical, and electromagnetic devices - usually silicon micromachined components operating in conjunction with computer controlled devices and radio transceivers to provide precision repeatability functions (such as in robotics artificial intelligence systems) emergency warning validation performance reconstruction.
- Computer vision –
    - Machine vision –
- Ubiquitous robotics – integrating robotic technologies with technologies from the fields of ubiquitous and pervasive computing, sensor networks, and ambient intelligence.

Contributing Fields

Robotics incorporates aspects of many disciplines including electronics, engineering, mechanics, software and arts. The design and control of robots relies on many fields knowledge, including:

- General
    - Aerospace –
    - Biology –
        - Biomechanics –
    - Computer science –
        - Artificial Intelligence –
        - Computational linguistics –
        - Cloud computing –
        - Cybernetics –
        - Modal logic –
    - Engineering –
        - Acoustical engineering –
        - Automotive engineering –
        - Chemical engineering –
        - Control engineering –
        - Electrical engineering –
        - Electronic engineering –
        - Mechanical engineering –
        - Mechatronics engineering –
        - Microelectromechanical engineering –
        - Nanoengineering –
        - Optical engineering –
        - Safety engineering –

        - Software engineering –
        - Telecommunications –
    - Fiction – Robotics technology and its implications are major themes in science fiction and have provided inspiration for robotics development and cause for ethical concerns. Robots are portrayed in short stories and novels, in movies, in TV shows, in theatrical productions, in web based media, in computer games, and in comic books. *See List of fictional robots and androids.*
        - Film – *See Robots in film.*
        - Literature – fictional autonomous artificial servants have a long history in human culture. Today's most pervasive trope of robots, developing self-awareness and rebelling against their creators, dates only from the early 20th century. *See Robots in literature.*
        - The Three Laws of Robotics in popular culture
    - Military science –
    - Psychology –
        - Cognitive science –
        - Behavioral science –
    - Philosophy –
        - Ethics –
    - Physics –
        - Dynamics –
        - Kinematics –
- Fields of application – additionally, contributing fields include the specific field(s) a particular robot is being designed for. Expertise in surgical procedures and anatomy, for instance would be required for designing robotic surgery applications.

Related Fields

- Building automation –
    - Home automation –
- Assistive technology
- Cloud robotics

ROBOTS

Types of Robots

Autonomous robot- robots that are not controlled by humans.

- Aerobot – robot capable of independent flight on other planets.
- Android – humanoid robot. Robot resembling the shape or form of a human.
- Automaton – early self-operating robot, performing exactly the same actions, over and over.
- Autonomous vehicle – vehicle equipped with an autopilot system, which is capable of driving from one point to another without input from a human operator.
- Ballbot – dynamically-stable mobile robot designed to balance on a single spherical wheel (i.e., a ball).
- Cyborg – also known as a cybernetic organism, a being with both biological and artificial (e.g. electronic, mechanical or robotic) parts.
- Explosive ordnance disposal robot – mobile robot designed to assess whether an object contains explosives; some carry detonators that can be deposited at the object and activated after the robot withdraws.
- Gynoid – humanoid robot designed to look like a human female.
- Hexapod (walker) – A six-legged walking robot, using a simple insect-like locomotion.
- Industrial robot – reprogrammable, multifunctional manipulator designed to move material, parts, tools, or specialized devices through variable programmed motions for the performance of a variety of tasks.
    - 3D printer
- Insect robot – small robot designed to imitate insect behaviors rather than complex human behaviors.
- Mobile robot – self-propelled and self-contained robot that is capable of moving over a mechanically unconstrained course.
- Military robot –
    - Cruise missile – robot-controlled guided missile that carries an explosive payload.
- Music Entertainment robot – robot created to perform music entertainment by playing custom made instrument or human developed instruments.
- Prosthetic robot – programmable manipulator or device replacing a missing human limb.
- Service robot – machines that extend human capabilities.
- Snakebot – robot or robotic component resembling a tentacle or elephant's trunk,

where many small actuators are used to allow continuous curved motion of a robot component, with many degrees of freedom. This is usually applied to snake-arm robots, which use this as a flexible manipulator. A rarer application is the snakebot, where the entire robot is mobile and snake-like, so as to gain access through narrow spaces.

- Surgical robot – remote manipulator used for keyhole surgery
- Walking robot – robot capable of locomotion by walking. Owing to the difficulties of balance, two-legged walking robots have so far been rare and most walking robots have used insect-like multilegged walking gaits.
- microbot- microscopic robots designed to go into the human body and cure diseases.
- nanobot - same as a microbot, just smaller. The components are at or close to the scale of a nanometer (10^{-9} meters).
- Rover (space exploration) -a robot with wheels designed to walk on other planets floors.

By Mode of Locomotion

Mobile robots may be classified by:

- The environment in which they travel:
    - Land or home robots. They are most commonly wheeled, but also include legged robots with two or more legs (humanoid, or resembling animals or insects).
    - Aerial robots are usually referred to as unmanned aerial vehicles (UAVs)
    - Underwater robots are usually called autonomous underwater vehicles (AUVs)
    - Polar robots, designed to navigate icy, crevasse filled environments
- The device they use to move, mainly:
    - Legged robot : human-like legs (i.e. an android) or animal-like legs.
    - Wheeled robot.
    - Tracks.

Robot Components and Design Features

- Actuator – motor that translates control signals into mechanical movement. The control signals are usually electrical but may, more rarely, be pneumatic or hydraulic. The power supply may likewise be any of these. It is common for electrical control to be used to modulate a high-power pneumatic or hydraulic motor.

- Linear actuator – form of motor that generates a linear movement directly.

■ Delta robot – tripod linkage, used to construct fast-acting manipulators with a wide range of movement.

■ Drive Power – energy source or sources for the robot actuators.

■ End-effector – accessory device or tool specifically designed for attachment to the robot wrist or tool mounting plate to enable the robot to perform its intended task. (Examples may include gripper, spot-weld gun, arc-weld gun, spray- paint gun, or any other application tools.)

■ Forward chaining – process in which events or received data are considered by an entity to intelligently adapt its behavior.

■ Haptic – tactile feedback technology using the operator's sense of touch. Also sometimes applied to robot manipulators with their own touch sensitivity.

■ Hexapod (platform) – movable platform using six linear actuators. Often used in flight simulators and fairground rides, they also have applications as a robotic manipulator.

See Stewart platform

■ Hydraulics – control of mechanical force and movement, generated by the application of liquid under pressure. *c.f. pneumatics.*

■ Kalman filter – mathematical technique to estimate the value of a sensor measurement, from a series of intermittent and noisy values.

■ Klann linkage – simple linkage for walking robots.

■ Manipulator – *gripper*. A robotic 'hand'.

  - Parallel manipulator – articulated robot or manipulator based on a number of kinematic chains, actuators and joints, in parallel. *c.f. serial manipulator.*

  - Remote manipulator – manipulator under direct human control, often used for work with hazardous materials.

  - Serial manipulator – articulated robot or manipulator with a *single* series kinematic chain of actuators. *c.f. parallel manipulator.*

■ Muting – deactivation of a presence-sensing safeguarding device during a portion of the robot cycle.

■ Pendant – Any portable control device that permits an operator to control the robot from within the restricted envelope (space) of the robot.

■ Pneumatics – control of mechanical force and movement, generated by the application of compressed gas. *c.f. hydraulics.*

■ Servo – motor that moves to and maintains a set position under command, rather than continuously moving.

■ Servomechanism – automatic device that uses error-sensing negative feedback to correct the performance of a mechanism.

- Single Point of Control – ability to operate the robot such that initiation or robot motion from one source of control is possible only from that source and cannot be overridden from another source.
- Slow Speed Control – mode of robot motion control where the velocity of the robot is limited to allow persons sufficient time either to withdraw the hazardous motion or stop the robot.
- Stepper motor –
- Stewart platform – movable platform using six linear actuators, hence also known as a Hexapod.
- Subsumption architecture – robot architecture that uses a modular, bottom-up design beginning with the least complex behavioral tasks.
- Teach Mode – control state that allows the generation and storage of positional data points effected by moving the robot arm through a path of intended motions.

Specific Robots

- Aura (satellite) – robotic spacecraft launched by NASA in 2004 which collects atmospheric data from Earth.
- Chandra X-ray Observatory – robotic spacecraft launched by NASA in 1999 to collect astronomical data.
- Robonaut – development project conducted by NASA to create humanoid robots capable of using space tools and working in similar environments to suited astronauts.
- Unimate – the first off-the-shelf industrial robot, of 1961.
- Justin (robot) –

Real Robots by Region

Robots from Australia

- GuRoo –
- UWA Telerobot –

Robots from Britain

- Black Knight
- eSTAR
- Freddy II
- George
- Shadow Hand
- Silver Swan

- Talisman UUV
- Wheelbarrow
- Robop

Robots from Canada

- Canadarm2 –
- Dextre –
- hitchBOT –
- ANATROLLER ARI-100 –
- ANATROLLER ARE-100 –
- ANATROLLER ARI-50 –
- ANATROLLER Dusty Duct Destroyer –
- ANAT AMI-100 –

Robots from China

- FemiSapien –
- Meinü robot –
- RoboSapien –
- Robosapien v2 –
- RS Media –
- Xianxingzhe –

Robots from Croatia

- DOK-ING EOD –
- TIOSS –

Robots from Czech Republic

- SyRoTek –

Robots from France

- Air-Cobot, a collaborative mobile robot able to inspect aircraft during maintenance operations –
- Digesting Duck –
- Jessiko –
- Nabaztag –
- Nao –

Robots from Germany

- BionicKangaroo – biomimetic robot model designed by Festo
- Care-Providing Robot FRIEND –
- Marvin –
- LAURON –

Robots from Italy

- IsaacRobot –
- Leonardo's robot –
- iCub –

Robots from Japan

- AIBO –
- ASIMO –
- Choromet –
- EMIEW –
- EMIEW 2 –
- Enon –
- Evolta –
- Gakutensoku –
- HAL 5v
- HOAP –
- KHR-1 –
- Omnibot –
- Plen –
- QRIO –
- R.O.B. –
- SCARA –
- Toyota Partner Robot –
- Wakamaru –

Robots from Mexico

- Don Cuco El Guapo –

Robots from the Netherlands

- Adelbrecht –
- Flame –
- Phobot –
- Senster –

Robots from New Zealand

- Trons, The –

Robots from Portugal

- RAPOSA –

Robots from Qatar

- Robot jockey –

Robots from Russia (or former Soviet Union)

- Lunokhod 1 –
- Lunokhod 2 –
- Teletank –

Robots from South Korea

- Albert Hubo –
- EveR-1 –
- HUBO –
- MAHRU –
- Musa –

Robots from Spain

- Maggie –
- REEM –
- Tico –

Robots from Switzerland

- Alice mobile robot –
- E-puck mobile robot –

- Pocketdelta robot –
- Shameer shami robot –

Robots from the United States of America

- Albert One –
- Allen –
- ATHLETE –
- Atlas –
- Baxter –
- Ballbot –
- avbotz Baracuda XIV –
- Beer Launching Fridge –
- Berkeley Lower Extremity Exoskeleton –
- BigDog –
- Boe-Bot –
- Coco –
- Cog –
- Crusher –
- Dragon Runner –
- EATR –
- Elektro –
- Entomopter –
- Haile –
- Hardiman –
- HERO –
- Johns Hopkins Beast –
- Kismet –
- Leonardo –
- LOPES –
- LORAX –
- Nomad 200 –
- Nomad rover –
- Octobot (robot) –
- Opportunity rover –
- Programmable Universal Machine for Assembly –
- Push the Talking Trash Can –

- RB5X –
- Robonaut –
- Shakey the Robot –
- Sojourner –
- Spirit rover –
- Turtle –
- Unimate –
- Zoë –
- Pleo –

Robots from Vietnam

- TOPIO –

International Robots

- European Robotic Arm –
- Curiosity Rover for NASA on Mars Science Laboratory space mission –

From British Radio

- Marvin the Paranoid Android *(Douglas Adams)* –

From British Television

- Kryten *(Rob Grant, Doug Naylor, David Ross, Robert Llewellyn)* {Red Dwarf} –
- Talkie Toaster – *(Rob Grant, Doug Naylor, John Lenahan, David Ross)* {Red Dwarf}
- K-9 (Doctor Who) –
- Robotboy – (*Bob Camp,* Charlie Bean, Heath Kenny, Prof Moshimo, Laurence Bouvard*) {Robotboy}*
- K.T., Eric and Desiree in Robert's Robots

Fictional Robots From the CZECH Republich

From Czech Plays

- Daemon – *(Karel apek)* {R.U.R. (Rossum's Universal Robots)}
- Helena – *(Karel apek)* {R.U.R. (Rossum's Universal Robots)}
- Marius – *(Karel apek)* {R.U.R. (Rossum's Universal Robots)}
- Primus – *(Karel apek)* {R.U.R. (Rossum's Universal Robots)}

- Radius – *(Karel apek)* {R.U.R. (Rossum's Universal Robots)}
- Sulla – *(Karel apek)* {R.U.R. (Rossum's Universal Robots)}

Fictional Robots From France

From French Ballets

- Coppélia – *(Arthur Saint-Leon, Léo Delibes)* {Coppélia}

From French Literature

- Hadaly – *(Auguste Villiers de l'Isle-Adam)* {The Future Eve}

Fictional Robots From Germany

From German Film

- Maschinenmensch – *(Fritz Lang, Thea von Harbou, Brigitte Helm)* {Metropolis}

From German Literature

- Maschinenmensch – *(Thea von Harbou)*
- Olimpia – *(E. T. A. Hoffmann)* {Der Sandmann}

Fictional Robots from Japan

From Anime

- Braiger – *(Shigeo Tsubota, Tokichi Aoki)* {Ginga Senpuu Braiger}
- Combattler V – *(Tadao Nagahama, Saburo Yatsude)* {Super Electromagnetic Robo Combattler V}
- Daimos – *(Tadao Nagahama, Saburo Yatsude)* {Brave Leader Daimos}
- Groizer X – *(Go Nagai)* {Groizer X}
- Mechander Robo – *(Jaruhiko Kaido)* {Mechander Robo (Gasshin Sentai Mekandaa Robo)}
- Raideen – *(Yoshiyuki Tomino, Tadao Nagahama)* {Brave Raideen}
- Trider G7 – *(Hajime Yatate)* {Invincible Robo Trider G7}
- Voltes V – *(Tadao Nagahama, Saburo Yatsude)* {Super Electromagnetic Machine Voltes V}

From Manga

- Astro Boy – *(Osamu Tezuka)* {Astro Boy}
- Doraemon – *(Fujiko Fujio)* {Doraemon}
- Getter Robo – *(Go Nagai, Ken Ishikawa)* {Getter Robo}
- Grendizer – *(Go Nagai)* {UFO Robo Grendizer}
- Mazinger Z – *(Go Nagai)* {Mazinger Z}
- Tetsujin 28 – *(Mitsuteru Yokoyama)* {Tetsujin 28 - Go!}

Fictional Robots From the United States of America

From American Comics

- Amazo – *(Gardner Fox)* {DC Comics}
- Annihilants – *(Alex Raymond)* {Flash Gordon}

From American Film

- C-3PO – *(George Lucas, Anthony Daniels)* {Star Wars}
- ED-209 – *(Paul Verhoeven, Craig Hayes, Phil Tippett)* {RoboCop}
- Fix-Its – *(Burton Weinstein, Robert Cooper, Tony Hudson)* {*batteries not included}
- Gort – *(Robert Wise, Harry Bates, Edmund H. North, Lock Martin)* {The Day the Earth Stood Still}
- Johnny Five – *(Tim Blaney, Syd Mead)* {Short Circuit}
- R2-D2 – *(George Lucas, Kenny Baker, Ben Burtt)* {Star Wars}
- Robby the Robot – *(Fred M. Wilcox, Robert Kinoshita, Frankie Darro, Marvin Miller)* {Forbidden Planet}
- The Terminator – *(James Cameron, Gale Anne Hurd)* {The Terminator}

From American Literature

- Adam Link – *(Eando Binder)* {I, Robot}
- Gnut – *(Harry Bates)* {Farewell to the Master}
- Robbie – *(Isaac Asimov)* {I, Robot}
- The Steam Man of the Prairies – *(Edward S. Ellis)* {The Steam Man of the Prairies}
- Tik-Tok – *(L. Frank Baum)* {Ozma of Oz}

From American Television

- Bender Bending Rodriguez – *(Matt Groening, David X. Cohen, John DiMaggio)* {Futurama}
- Cambot – Gypsy, Crow T. Robot, and Tom Servo *(Joel Hodgson, Trace Beaulieu, Bill Corbett, Josh Weinstein, Jim Mallon, Patrick Brantseg)* {Mystery Science Theater 3000}
- Data – *(Gene Roddenberry, Brent Spiner)* {Star Trek: The Next Generation}
- GIR – *(Jhonen Vasquez, Rosearik Rikki Simons) {Invader Zim}*
- Jenny Wakeman – *(Rob Renzetti, Janice Kawaye) {My Life as a Teenage Robot}*
- Robot B-9 – *(Irwin Allen, Robert Kinoshita, Bob May, Dick Tufeld)* {Lost in Space}
- Grounder and Scratch – *(Phil Hayes, Garry Chalk*) {Adventures of Sonic the Hedgehog*}*

REFERENCES

1. "Cloud Robotics and Automation A special issue of the IEEE Transactions on Automation Science and Engineering.". IEEE. Retrieved 7 December 2014.
2. "RoboEarth".
3. Goldberg, Ken. "Cloud Robotics and Automation".
4. Li, R. "Cloud Robotics-Enable cloud computing for robots". Retrieved 7 December 2014.
5. Kehoe, Ben; Patil, Sachin; Abbeel, Pieter; Goldburg, Ken (13 September 2014). "A Survey of Research on Cloud Robotics and Automation" (PDF). IEEE TRANSACTIONS ON AUTOMATION SCIENCE AND ENGINEERING.
6. "Impact of Cloud Computing on Healthcare" (PDF).
7. Li, Ruijiao; Hu, Huosheng (16 October 2013). "Towards ROS Based Multi-robot Architecture for Ambient Assisted Living". Systems, Man, and Cybernetics (SMC), 2013 IEEE International Conference on: 3458–3463. doi:10.1109/SMC.2013.590.
8. "Project of the Future: Industry 4.0". Retrieved 9 December 2014.
9. LaSelle, Rush. "Automation in the Cloud". Robotic Industries Association. Retrieved 9 December 2014.
10. "roboearth". Retrieved 7 December 2014.
11. Waibel, M; Tenorth, M; D'Andrea, R. "RoboEarth". IEEE Robotics & Automation Magazine. 18 (2): 69–82. doi:10.1109/MRA.2011.941632.
12. "Rapyuta". Retrieved 7 December 2014.
13. Hunziker, D; D'Andrea, R; Gajamohan, M; Waibel, M (May 2013). "Rapyuta: The RoboEarth Cloud Engine". Robotics and Automation (ICRA), 2013 IEEE International Conference on: 438–444. doi:10.1109/ICRA.2013.6630612.

14. "KnowRob". Retrieved 8 December 2014.
15. "RoboBrain Project". Retrieved 7 December 2014.
16. "Robo Brain' mines the Internet to teach robots".
17. "MyRobots". Retrieved 9 December 2014.
18. Hu, Huosheng; McDonald-Maier, Klaus D; Gu, Dongbing; Li, Ruijiao. "COLAS". Retrieved 7 December 2014.
19. "ROSjava-Cloud Robotics". Retrieved 9 December 2014.
20. Robotics-vo. "A Roadmap for U.S. Robotics From Internet to Robotics 2013 Edition" (PDF). Retrieved 8 December 2014.
21. "2013 IEEE IROS Workshop on Cloud Robotics. Tokyo. November 2013".
22. "http://cloud-robotics.cs.umn.edu/". Retrieved 7 December 2014.External link in |title= (help)
23. "RobotApp". Retrieved 7 December 2014.
24. "DARPA-Cloud-Robotics". Retrieved 7 December 2014.
25. Goldberg, Ken. "Cloud-based robot grasping with the google object recognition engine". Robotics and Automation (ICRA), 2013 IEEE International Conference on. doi:10.1109/ICRA.2013.6631180.
26. "Big NSF grant funds research into training robots to work with humans".
27. Kuffner, James. "Cloud Enabled Humanoid Robots" (PDF). IEEE-RAS International Conference on Humanoid Robotics.
28. "Google I/O 2011: Cloud Robotics, ROS for Java and Android". Retrieved 9 December 2014.
29. "Networked Robots Committee". Retrieved 8 December 2014.
30. Dautenhahn, Kerstin (29 April 2007). "Socially intelligent robots: dimensions of human–robot interaction" (pdf). Phil. Trans. R. Soc. b. 362 (1480): 679–704. doi:10.1098/rstb.2006.2004.
31. Mathur, Maya B.; Reichling, David B. (2016). "Navigating a social world with robot partners: a quantitative cartography of the Uncanny Valley" (PDF). Cognition. 146: 22-32. doi:10.1016/j.cognition.2015.09.008.
32. Fong, T.; Kunz, C.; Hiatt, L.; Bugjska, M. (2006). "The Human-Robot Interaction Operating System" (PDF).
33. Casper, J.; Murphy, R. (June 2003). "Human-robot interactions during the robot-assisted urban search and rescue response at the World Trade Center" (PDF). IEEE Transactions on Systems, Man, and Cybernetics. 33 (3): 367–385. doi:10.1109/tsmcb.2003.811794.
34. Bartneck, Christoph; Michio Okada (2001). "Robotic User Interfaces" (PDF). Proceedings of the Human and Computer Conference. pp. 130–140.
35. Bartneck, Christoph (February 2011). "The end of the beginning: a reflection on the first five years of the HRI conference". Scientometrics. 86 (2): 487–504. doi:10.1007/s11192-010-0281-x.

36. According to the Oxford English Dictionary, the term "robotics" was first used in the short story "Liar!" published in the May, 1941 issue of Astounding Science Fiction.
37. V. Daniel Hunt (1983), "Appendix A - Glossary", Industrial robotics handbook, Industrial Press Inc., ISBN 978-0-8311-1148-9
38. Helena Domaine (2006), "Glossary", Robotics, Lerner Publications, ISBN 978-0-8225-2112-9
39. Joseph A. Angelo (2007). Robotics: a reference guide to the new technology. Libraries Unlimited. pp. 258–327. ISBN 978-1-57356-337-6. Retrieved 28 January 2011.
40. "OSHA Technical Manual - SECTION IV: CHAPTER 4 - INDUSTRIAL ROBOTS AND ROBOT SYSTEM SAFETY". Occupational Safety and Health Administration. Retrieved 2011-01-28.
41. Rail track and Linear track (PDF)
42. "DARPA ROBOTICS CHALLENGE (DRC)". Retrieved 14 January 2013.

Index

M

P

R

S

T

U

V

W